678·675

Rocket Propulsion Establishment
Library

Please return this publication, or request renewal, before
the last date stamped below.

Name	Date

RPE Form 243 (revised 6/71)   739490

243

ACADEMY OF SCIENCES OF THE USSR
Institute for High Molecular Compounds

N.A. Adrova, M.I. Bessonov,
L.A. Laius, and A.P. Rudakov

POLYIMIDES

A NEW CLASS
OF HEAT-RESISTANT POLYMERS

Translated from Russian by J. Schmorak

Israel Program for Scientific Translations
Jerusalem 1969

Distributed by:
ann arbor-humphrey
science publishers ltd.
5 Great Russell St., London, W C 1.

This book is a translation of

POLIIMIDY — NOVYI KLASS
TERMOSTOIKIKH POLIMEROV

Izdatel'stvo "Nauka"
Leningrad 1968

IPST Cat. No. 2211

Printed and Bound in Israel
Printed in Jerusalem by IPST Press
Binding: Wiener Bindery Ltd., Jerusalem

Table of Contents

INTRODUCTION

High-molecular compounds have now become the starting materials in the preparation of a large number of technical products: plastic masses, films, resins, fibers, lacquers etc., the quality of which determines the service performance of a large number of details and constructions used for various purposes. The technical specifications which these materials have to meet often determine the course of purely scientific studies of physics and chemistry of high polymers. One such requirement is that the polymeric materials withstand high temperatures without impairment of their characteristic advantages and specific properties. This is particularly important in electrical engineering, power engineering and aviation engineering, in which the effectiveness of numerous devices depends on their maximum service temperature. For this reason, the production of heat-resistant polymers, i.e., polymers combining temperature resistance* and thermal stability,** has been for many years one of the main tasks of chemistry of high-molecular compounds. The temperature resistance of high polymers is mainly determined by the intensity of intermolecular forces, presence of cross-linking, crystallinity, etc. Thermal stability is determined mainly by the resistance to heat of individual groups of the macromolecule and of the bonds by which they are linked into a macromolecular chain. Only in isolated cases — mainly by analogy with model compounds and on the strength of results obtained by studying homologous polymer series — is it possible to predict the temperature resistance and thermal stability of a given polymer in advance, from its molecular structure alone. Numerical values of temperature resistance and thermal stability are obtained by determining certain physical properties, which are connected in a complex manner with the chemical structure of the polymer and which are sometimes selected in a highly arbitrary manner. Finally, from the point of view of its service performance, the polymer must have superior physical and mechanical properties and should be easily processed. It is clear, accordingly, that the development of new heat-resistant polymers with a superior service performance still remains a very exacting and complicated task, which includes not only a wide range of syntheses, but also a detailed study of different properties of high polymers and high-polymeric materials in a wide range of temperatures.

For many years the temperature resistance and thermal stability of high polymers were enhanced by modifying the structure of the side groups, while retaining the main carbon chain unchanged, or by introducing heat-resistant groups, such as rings, into the chain. As a result of these studies there appeared such well known polymeric materials as polytetrafluoroethylene (Teflon) and polyethyleneterephthalate (Terylene, Lavsan).

The next step was the preparation of fully cyclic polymers which do not contain aliphatic links in the chain. The melting and softening points of

---

* As characterized by the value of the melting point, softening point, etc. Indicates the maximum temperature at which the polymer can still be utilized as solid.

** As characterized by the time during which given values of mechanical strength and other properties are retained at high temperatures. Indicates the practical working temperature range.

such compounds are of the order of 300—400°, while decomposition occurs at even higher temperatures. Their technological application was impeded owing to their poor mechanical and physical processes and owing to the fact that they were difficult and sometimes impossible to process. Recently, however, new synthetic methods have been developed with the result that processing has become possible, and the physical and mechanical properties of the products have improved without impairment of their temperature resistance and thermal stability; this was achieved by selecting the proper combination of the cyclic links with the interlinking groups.

It may be said with certainty that one of the most successful accomplishments in this respect was the synthesis of polyimides — ring-chain polymers of the structure

$$\left[-N\underset{CO}{\overset{CO}{<}}R\underset{CO}{\overset{CO}{>}}N-R'- \right]_n,$$

where R and R' are aromatic and other heat-resistant groupings. This is probably the first example of ring-chain polymers which are greatly superior to ordinary chain polymers not only as regards their temperature resistance and thermal stability, but also as regards the totality of their physical and mechanical properties and the variety of their different applications. The first polymers with imide links in the chain were obtained as early as the beginning of the present century, but they began to be extensively studied and utilized only in the late 1950's, after a two-step synthesis had been developed in the U.S.A. by DuPont. For a number of years this field of chemistry and polymer technology remained the monopoly of this company, which took out patents for the most important procedures of preparation of polyimides and polyimide materials and kept strict secrecy as to their methods of production. At present polyimides are studied and successfully produced by other major American companies. Extensive systematic work on the subject is also being conducted in the Soviet Union, Japan, Great Britain, West Germany and other countries.

The aim of the present book is to familiarize the reader with the most important methods of preparation of polyimides, their physical properties and service performance. It may be assumed that this class of polymers will be extensively studied and utilized in the near future. The authors do not claim to have given an exhaustive treatment of the chemistry, physics and technology of these polymers, since the results of scientific investigations on the subject are still scanty, while technological data in the literature are practically nonexistent. The book is based mainly on the data published in scientific periodicals up to the end of 1966. The authors have taken the liberty of giving a detailed account of studies which in their view are of principal significance in the understanding of the nature of polyimides and the relationships governing their physicochemical transformations, while being fully aware that there is still much which has remained inconclusive and unclear. Accordingly, stress has been laid on the experimental data in the hope that these will be of help to the reader in his further studies and utilization of polyimides.

Data found in the patent literature available on the subject have been considered only to the extent to which they may be said to lead to conclusions of general importance. A selection of the most important patents with a brief indication of the nature of each invention is also given.

Chapter I

SYNTHESIS AND TRANSFORMATIONS OF POLYIMIDES

Polyimides are ring-chain polymers. The first synthesis of polyimides was accomplished by Bogert and Renshaw /40/ who noted that when 4-aminophthalic anhydride or dimethyl 4-aminophthalate were heated, water and methanol, respectively, were evolved with the formation of polyimide:

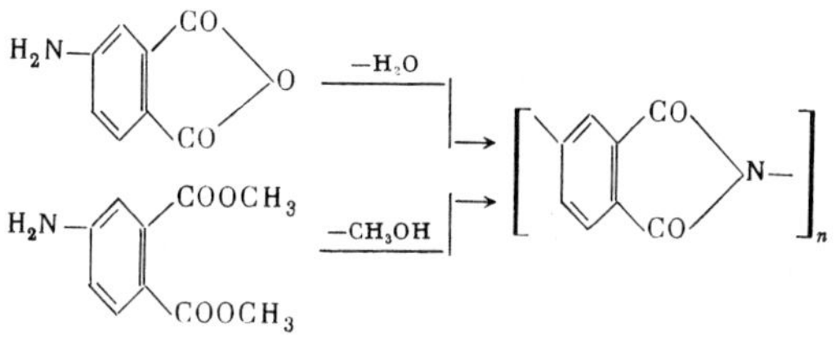

However, special interest in the synthesis of polyimides arose only in recent years when it became clear that many polymers of this class display extremely valuable properties.

As regards their structure and preparation techniques, polyimides may be subdivided into two main groups:

1) polyimides with aliphatic links in the main chain;
2) polyimides with aromatic links in the main chain.*

Polyimides with aliphatic links in the main chain, of the general formula

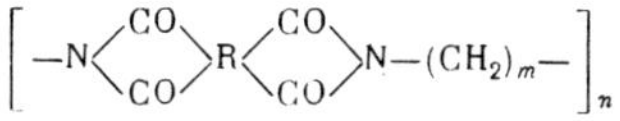

are prepared by thermal polycondensation by heating salts of aromatic tetracarboxylic acids and aliphatic diamines.

Polyimides with aromatic links in the main chain, of the general formula

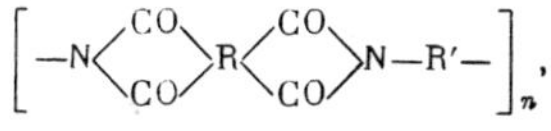

, where $R' = Ar$,

are prepared, as a rule, by two-step polycondensation. This technique is now extensively employed on account of the fact that its first step yields soluble products. The first reaction stage consists in the acylation of the

---

* Briefly referred to as polyarimides /20/. This designation will be used in the present book.

diamine by the tetracarboxylic acid dianhydride in a polar solvent, with the formation of polyamido-acid, according to the equation

$$O<\begin{smallmatrix}CO\\CO\end{smallmatrix}>R<\begin{smallmatrix}CO\\CO\end{smallmatrix}>O + H_2N-R'-NH_2 \longrightarrow$$

$$\longrightarrow \left[\begin{array}{c} HOOC \\ -NH-CO \end{array} >R< \begin{array}{c} CO-NH-R'- \\ COOH \end{array} \right]_n .$$

The second step of the synthesis — cyclodehydration of the polyamido-acids (imidization) — proceeds according to the equation

$$\left[\begin{array}{c} HOOC \\ -NH-CO \end{array} >R< \begin{array}{c} CO-NH-R'- \\ COOH \end{array} \right]_n \xrightarrow{-2nH_2O}$$

$$\longrightarrow \left[-N< \begin{smallmatrix}CO\\CO\end{smallmatrix} >R< \begin{smallmatrix}CO\\CO\end{smallmatrix} >N-R'- \right]_n .$$

and may be effected thermally or chemically.

In a recent method, aromatic polyimides are prepared by the interaction of diimides and dihalides in solution:

$$HN< \begin{smallmatrix}CO\\CO\end{smallmatrix} >R< \begin{smallmatrix}CO\\CO\end{smallmatrix} >NH + X-R'-X \xrightarrow{-HX}$$

$$\longrightarrow \left[-N< \begin{smallmatrix}CO\\CO\end{smallmatrix} >R< \begin{smallmatrix}CO\\CO\end{smallmatrix} >N-R'- \right]_n .$$

Preparation of polyimides by polycondensation in the melt

Edwards and Robinson /61/ were the first to describe a preparation of an aliphatic polyimide by melting a diamine with a tetracarboxylic acid. This reaction can be represented as follows:

$$\begin{array}{c} HOOC \quad COOH \\ \text{(ring)} \\ CH_3OOC \quad COOCH_3 \end{array} + H_2N-(CH_2)_m-NH_2 \xrightarrow{\Delta}$$

$$\longrightarrow \begin{array}{c} HOOC \quad COO^{\ominus} \overset{\oplus}{NH_3}-(CH_2)_m-NH_2 \\ \text{(ring)} \\ CH_3OOC \quad COOCH_3 \end{array} \xrightarrow{\Delta}$$

$$\longrightarrow \left[-N< \begin{smallmatrix}CO\\CO\end{smallmatrix} >\text{(ring)}< \begin{smallmatrix}CO\\CO\end{smallmatrix} >N-(CH_2)_m- \right]_n .$$

After heating at 110—138° the low-molecular intermediate product (salt) is formed. It is converted to the polyimide by several hours additional heating at 250—300°.

The technique of polycondensation in the melt is of only limited application in the preparation of polyimides. The melting point of the resulting polyimide must be below the reaction temperature, so as to

2

obtain a molten reaction product during polycondensation; this is a necessary condition for obtaining products with a high molecular weight. For this reason melt polycondensation can be successfully employed only if the aliphatic diamine contains not less than seven methylene groups. Aromatic diamines are not basic enought to form salts with carboxylic acids. In addition, aromatic polyimides are as a rule infusible, so that the reaction product solidifies before a sufficiently high molecular weight can be attained.

The softening temperature of aliphatic polyimides depends on the structure of the diamines. Table 1 shows the relevant properties of polypyromellitimides (polyimides based on pyromellitic acid dianhydride) obtained in the melt, using various aliphatic diamines.*

TABLE 1. Properties of aliphatic polypyromellitimides /99/

Diamine	Stability to oxidation at 175°C, hours	Vitrification temperature, °C
Nonamethylene diamine	20—25	110
4,4'-Dimethylheptamethylene diamine	20—30	135
3-Methylheptamethylene diamine	8—10	135

The utilization of diamines with a large number of methylene groups, such as, say, 2,11-diaminododecane, yields polypyromellitimides with a melting temperature below 300°. Polypyromellitimide from 4,4'-dimethylheptamethylene diamine has a melting temperature of 320°. A polymer with η_{log} = 0.5 (0.5% solution in m-cresol) and a melting point of 325° has been obtained /67/ from nonamethylene diamine. The melt of this polymer may be drawn into fibers.

Polyimides capable of melting without decomposition were also prepared /99/ from other dianhydrides of the general formula

where R = —O—; —CH₂—, and from diamines with shorter chains, for example hexamethylene diamine and tetramethylene diamine.

Polycondensation in the melt has been employed /106/ to prepare polyimides from alicyclic tetracarboxylic acid dianhydrides — cyclobutane-tetracarboxylic acid dianhydride (I) and the dianhydride of tricyclo-(4,2,2,0²·⁵)-deca-10-ene-3,4,7,8-tetracarboxylic acid (II):

* Tables and figures not provided with literature references represent the authors' own data.

3

Upon treatment with methanol these dianhydrides were converted to half-esters of the corresponding tetracarboxylic acids, which in turn gave salts by reacting with diamines. The following diamines were employed: hexamethylene diamine, octamethylene diamine, nonamethylene diamine, decamethylene diamine, dodecamethylene diamine and 4,4'-diamino-diphenylmethane. Salt polycondensation was conducted both in the melt and in a solution of diphenyl ether. As distinct from pyromellitic and other aromatic dianhydrides, carboxylic groups of alicyclic dianhydrides are not located in the plane of the ring. Moreover, in this case the ring is more mobile, so that the imide bonds may be formed between the carboxyl groups of different macromolecules, which may result in cross-linking. Accordingly, when the dianhydride of tricyclodecaene-tetracarboxylic acid (II) was condensed with diamines, the results were more satisfactory in solution than in the melt. Polymers obtained with nonamethylene diamine and decamethylene diamine had softening temperatures of 230 and 255° and reduced viscosities of 0.45 (in dimethylformamide) and 1.4 (in tricresol) respectively. These polyimides are capable of film formation. Reaction between these diamines and the dianhydride of cyclobutane-tetracarboxylic acid (I) yielded only low-molecular polyimides.

Cross-linking in polypyromellitimides obtained by heating salts of pyromellitic acid and aliphatic diamines (hexamethylene diamine and nonamethylene diamine) was studied by Hermans and Street /70/. These authors found that m-cresol-soluble polynonamethylenepyromellitimide is obtained if the salt is heated under conditions preventing the elimination of the volatile products. Unless this is ensured, the reaction mixture evolves not only the reaction products but also pyromellitic acid dianhydride, the resulting excess of free terminal amino groups causes cross-linking, and the product is no longer soluble in m-cresol. If hexamethylene diamine is employed, the product is always insoluble.

Two-step preparation of aromatic polyimides

The first reports about the two-step polycondensation synthesis of polyimides appeared in 1959 in the patents of DuPont and then in several publications /10, 42, 66, 101, etc./. This method, which is now extensively employed, yields polyimides based on dianhydrides of tetracarboxylic acids of different classes and aromatic diamines. The general equations describing the fundamental reactions involved will be found on p. 2.

Synthesis and study of polyamido-acids. The first step in polyimide synthesis — preparation of polyamido-acid — is usually performed as follows. To a solution of the aromatic diamine in a suitable solvent an equimolecular amount (or a 1—5% excess) of the dry tetracarboxylic acid dianhydride is introduced in small portions, with stirring. As more dianhydride is added, the viscosity of the solution gradually increases. When the reagent ratio becomes nearly equimolecular, there is a sharp increase in the viscosity of the solution. The reaction is conducted between -20 and +70°. If the temperature is raised above 70°, the molecular weight of the polyamido-acid decreases. In most cases 15—20° is the optimum reaction temperature, at which a high-molecular polyamido-acid is obtained. The reaction is conducted in polar solvents which strongly associate with the reagents, forming, in particular, reactive complexes with the dianhydrides of tetracarboxylic acids. The best solvents for such reactions are N, N-dimethylacetamide, N, N-dimethylformamide, dimethyl sulfoxide and N-methyl 2-pyrrolidone. These solvents are used both alone and in combination with benzene, benzonitrile, dioxane, xylene, toluene, cyclohexane, etc.

The mechanism of formation of polyamido-acids and the effect of different factors on the course of this stage of polyimide preparation have been studied in most detail.

Figure 1 shows the variation in the viscosity of the reaction mixture in the preparation of the polyamido-acid during the interaction between pyromellitic dianhydride with 4, 4'-diaminodiphenylmethane in different solvents solvents /42/. It is seen that dimethyl sulfoxide, which is the most highly viscous solvent, yields a polyamido-acid with a higher molecular weight than do N, N-dimethylformamide and N, N -dimethylacetamide. The maximum viscosity is noted in all cases when the reagent ratio is somewhat higher than equimolecular. Large excess of the dianhydride yield a polyamido-acid of a much lower molecular weight.

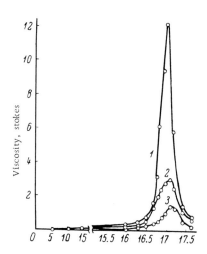

Amount of pyromellitic dianhydride, grams

FIGURE 1. Viscosity of polyamido-acid solution as a function of the amount of pyromellitic dianhydride introduced into the mixture, for a constant amount of 4.4'-diaminodiphenylmethane /42/.

1 — in dimethyl sulfoxide; 2 — in N, N-dimethylacetamide; 3 — in N, N-dimethylformamide.

Bower and Frost /42/ also investigated the effect of the sequence of addition of the reagents on the molecular weight of the polyamido-acid. They showed that the highest molecular weight is obtained when dry dianhydride of the tetracarboxylic acid is added in small portions to the solution of the diamine. If the addition sequence of the reagents is changed by, say, slowly adding the diamine to the solution of the dianhydride, or if the reagents are mixed together in equimolecular proportions, the viscosity of the solution will be too low. This may be due to the rupture of the cross-links in the polymer by dianhydrides if these are present in excess.

Frost and Kesse /66/ studied the effect of the ratio of the starting reagents on the variation in the viscosity of solutions of polyamido-acids during their formation and on prolonged standing for the case of pyromellitic dianhydride and 4, 4'-methylenedianiline. The reagent ratio varied between unity and 1.04. The results of these experiments are shown in Figure 2 and Table 2 from which it can be seen that the optimum dianhydride: diamine ratio is 1.015 — 1.020. If the reagents are taken in an equimolecular ratio or if the dianhydride is present in larger excess, the viscosity of the solutions is low.

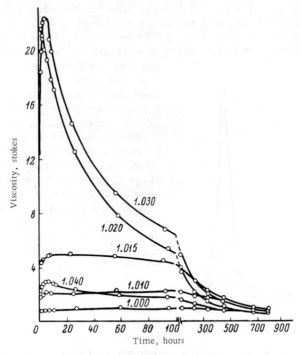

FIGURE 2. Effect of the pyromellitic dianhydride: 4,4'-diamino-diphenylmethane ratio (values of molar ratio shown on the curves) on the variation of the viscosity of 12% solution of polyamido-acid with time at 35°C /66/.

TABLE 2. Effect of reagent ratio on the viscosity of polymer solution obtained from pyromellitic dianhydride and 4,4'-methylenedianiline /66/

Molar ratio dianhydride : diamine	Maximum viscosity, stokes	Time during which maximum viscosity attained, minutes	Viscosity after 2140 hours at 35°C, stokes
1.000	0.946	60	0.227
1.010	2.195	58	0.369
1.015	4.95	12	0.372
1.020	21.91	0.3	0.357
1.030	22.39	4.2	0.309
1.040	2.892	4.2	0.316

It follows from /12/ that the specific viscosity of polyamido-acids also depends on the concentrations of the reagents in the solvent. There is an optimum concentration at which the resulting high polymer has a maximum viscosity.

Frost and Kesse /66/ showed that the viscosity of the polyamido-acid obtained from pyromellitic anhydride and 4,4'-diaminodiphenyl ether is affected by the presence of moisture. Their experiments were conducted on a 10% solution of the polymer in N, N-dimethylacetamide containing 0.02% (0.04 mole) of water per repeating unit of the polymer. Figure 3 shows the variation of viscosity with time in this practically anhydrous system as compared with experiments in which the nondried solvent (0.12% water) and solvent containing added water (4.4%) were employed. The viscosity decrease in the anhydrous system was much smaller than in systems containing some water. However, it was found during a study of the effect of water on the polycondensation of the dianhydride of 3, 3', 4, 4'-benzophenonetetracarboxylic acid with aromatic diamines /14/ that the presence of a small amount of water (about 1%) in dimethylformamide has the opposite effect of increasing the specific viscosity of the polyamido-acid if the reaction is conducted at 20° and the reactant concentration is 25%. The authors of /14/ consider that at 20° water is a polymerization catalyst, whereas at higher temperatures it brings about hydrolysis of the polyamido-acids, the rate of which increases with increasing temperature.

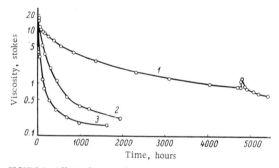

FIGURE 3. Effect of water of the temporal variation of viscosity of a 10% solution of a polyamido-acid based on pyromellitic dianhydride and 4,4'-diaminodiphenyl ether in dimethylacetamide at 35°C /66/.

1 — anhydrous; 2 — 0.12% water; 3 — 4.4% water.

Solutions of polyamido-acids may be stored at 0° for a long time without marked changes in the viscosity. The stability of the solutions largely depends on their concentration. The drop in viscosity is much more rapid in dilute than in concentrated solutions.

Most polyamido-acids yield transparent, faintly colored elastic films when their solutions are evaporated. It should be noted that when these films are redissolved, the resulting solutions are as a rule less viscous than the solutions of the initial polyamido-acids.

Polyamido-acids begin to soften at about 100—160°, depending on their structure. However, it would be difficult to determine the softening points of these polymers accurately, since they become dehydrated in the heat, forming the corresponding polyimides /4/. Moreover, polyamido-acids invariably contain a considerable amount of bound solvent.

The information available on the molecular weight of polyamido-acids is mostly fragmentary and is usually limited to the intrinsic viscosity value or, at best, to an estimate of the molecular weight with the aid of universal, but not always applicable formulas. A recent study /112/ reported the results of a detailed investigation of the molecular characteristics of a polyamido-acid prepared from pyromellitic dianhydride and 4,4'-diaminodiphenyl ether, with the repeating unit

Wallach /112/ studied 13 samples of polyamido-acids obtained by homogeneous and heterogeneous polycondensation of different samples of the dianhydride and the diamine in a solvent (N, N-dimethylacetamide) of different degrees of purity.

The large number of methods employed (viscometry, osmometry, light scattering, sedimentation rate) and the interpretation of the results obtained in the light of modern theoretical concepts furnished information not only on the molecular weight and molecular weight distribution, but also on the dimensions, flexibility and intramolecular interaction of the polyamido-acid macromolecules. Wallach determined the numerical values of the coefficients in the equations connecting the intrinsic viscosity $[\eta]$ of the polyamido-acid in a solution of purified dimethylacetamide containing LiBr and the sedimentation constant S_0 with the weight-average molecular weight of the polyamido-acid:

$$[\eta] = 2.38 \cdot 10^{-4} \cdot M_w^{0.78},$$
$$S_0 = 3.72 \cdot 10^{-2} \cdot M_w^{0.36}.$$

It is recommended that these equations be employed in the first instance in the case of heterogeneous polycondensation, when the solid dianhydride is added to the solution of the diamine. Table 3 shows some of the data reported in /112/ on the molecular weights of polyamido-acids prepared under various conditions.

It is seen from Table 3 that the polycondensation technique and the degree of purity of the monomers to a large extent determine the molecular weight and the molecular weight distribution of the product. The highest number-average molecular weight (M_n = 55,000) and the narrowest molecular weight distribution (M_w/M_n = 2.39) were obtained in the case of heterogeneous polycondensation of particularly pure reagents (sample 9). On the contrary, a wide molecular weight range is obtained if the solvent has not been purified (samples 2, 11, 13, for which $M_w/M_n > 3$) and when a solution of the diamine was added to the solution of the dianhydride (samples 6 and 12).

These data show that polyamido-acids have, as a rule, a low molecular weight (less than 100,000) and a low degree of polymerization (not above 140). Analysis of concentrational relationships involving viscosity, sedimentation and light scattering lead to the conclusion /112/ that polyamido-acid macromolecules are highly flexible (size of segment proved to be near the size of a monomeric unit) and that the interaction between side groups is weak, much weaker, for example, than in the case of cellulose derivatives. This is due to the presence of ether and other bonds, around which internal rotation is possible.

TABLE 3. Molecular characteristics of polyamido-acids based on pyromellitic dianhydride and 4,4'-diaminodiphenyl ether /112/

A. Conditions of polycondensation and purification

Sample	Polycondensation	Purification of		
		diamine	dianhydride	solvent
2	Heterogeneous	Vacuum drying	Vacuum drying	Not purified
11	"	"	"	"
13	"	"	Two recrystallizations	"
9	"	"	"	P
7	"	"	Sublimation	Distillation
6	Homogeneous	"	Two recrystallizations	P
12	"	"	"	P

No t e s. In heterogeneous polycondensation the reagent added was the dianhydride; in homogeneous polycondensation it was the dianhydride (sample 6) or the diamine (sample 12). P—distillation over P_2O_5 followed by holding over molecular sieve.

B. Molecular weight characteristics

Sample	$[\eta]$	S_0	$M_n \cdot 10^{-4}$	$M_w \quad 10^{-4}$	M_w/M_n
2	1.20	1.79	1.84	5.72	3.11
11	2.76	2.73	4.98	19.2	3.86
13	4.00	3.22	4.54	26.6	4.80
9	2.40	3.05	5.49	13.1	2.39
7	1.94	2.11	2.49	6.82	2.74
6	1.85	2.32	3.73	9.52	2.55
12	2.76	2.79	3.96	14.3	3.61

The synthesis of a polyamido-acid is a bimolecular acylation of amineś, consisting in a nucleophilic attack on the amino group accompanied by the opening of the anhydride ring and the formation of the polymer. The mechanism of aminolysis of carboxylic acid anhydrides by monofunctional amines was studied by UV spectroscopy /76/. The mechanism of the formation of the polyamido-acid by the reaction between pyromellitic dianhydride and tetramethyl-p-phenylene diamine was studied by EPR /2/. Since pyromellitic dianhydride is an electron acceptor, whereas aromatic diamines have a low ionization potential, complexes may be formed with a charge transfer between the reacting compounds. A typical indication of complex formation is the intense coloration which appears when pyromellitic dianhydride is added to the diamine solution. However, the color rapidly disappears, since the complex formation is accompanied by a rapid proton transfer from the amine to the anhydride group and formation of the amide bond:

Wrasidlo et al. /113/ studied the mechanism and kinetics of the reaction between pyromellitic dianhydride with m-phenylene diamine by IR spectroscopy:

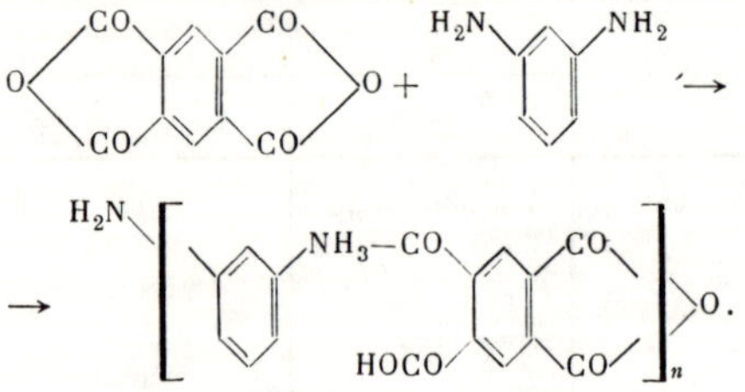

Equilibrium in the system is attained when 78 mole % of the dianhydride have reacted. The free amine residual in the reaction mixture may react with the formation of "nylon" type salt

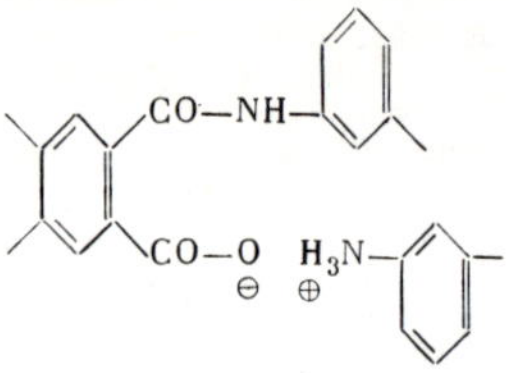

This salt may react at elevated temperatures to yield the corresponding amides, which in turn form the polyimide and polymeric chains with terminal amino groups:

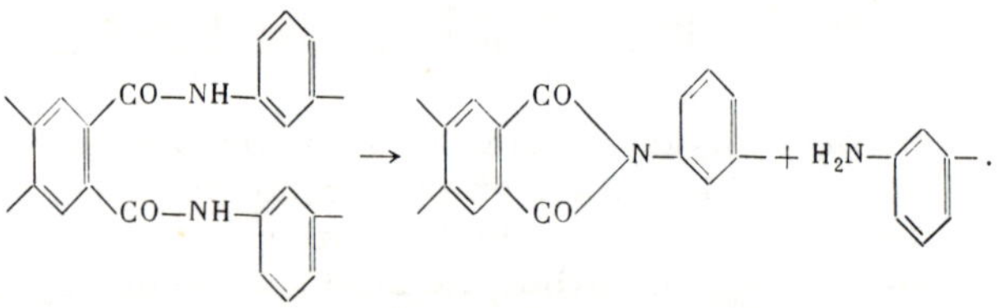

The effect of a large excess of dianhydride and diamine on the mechanism of polymer formation was studied by Frost and Kesse /66/ for the reaction between pyromellitic acid dianhydride and 4, 4'-diaminodiphenyl ether in dimethylacetamide solution. A 25% excess of each monomer was taken in the experiment. It was found that an excess of both the diamine and the dianhydride results in a rapid degradation. Similar results were obtained when monofunctional reagents — phthalic anhydride and aniline — were employed. This effect may be explained /66/ by the redistribution reactions taking place in the system which are responsible for the establishment of the equilibrium involving a constant molecular weight and a stable viscosity.

If the component ratio is equimolecular, the reaction between pyromellitic dianhydride and the diamine is more rapid. As the concentration of the anhydride and amino groups decreases, the reaction rate tends to zero. In the absence of all other reactions, the molecular weight of the polyamido-acid would approach a certain limiting value, determined by the ratio of the initial components. However, as the rate of the main reaction decreases, side reactions assume a major importance. The scheme /66/ which includes all the processes in the system is given below:

$$
\text{(structure I)} \longrightarrow \text{(structure)} \; N- + H_2O \qquad (1)
$$

$$
\text{(II)} \rightleftarrows \; \rightleftarrows \; \longrightarrow
$$

$$
\longrightarrow \; O + H_2O \qquad (2)
$$

$$
O + H_2N- \xrightarrow{H_2O} \; \begin{array}{c} COOH \\ COOH. \end{array} \qquad (3)
$$

According to this scheme, structure I is intermediate in the main imidization reaction (1) which at room temperature is slow, irreversible, and becomes the main reaction after a long time has elapsed. Another process which takes place in the system is represented by equation (2) and is a second irreversible reaction (isoimide formation). Structure II evolves water, isoimide and an amino group with the formation of the anhydride, which in the presence of water forms a dicarboxylic acid according to reaction (3). In the early stage, in a practically anhydrous medium, this degradation reaction is insignificant, but in the presence of water it proceeds at a rapid rate according to mechanism (3) from the very beginning. As the degradation proceeds, the polymer decomposes to short chains with terminal carboxyl and amino groups.

In addition to all the above factors, the course of formation of the polyamido-acid depends on the nature of the reacting substances. Thus, it was shown /13/ by a study of the polycondensation of pyromellitic dianhydride with different diamines that amines may be arranged in the following activity sequence in the reaction of polyamido-acid formation: hexamethylene diamine > decamethylene diamine > 4, 4'-diaminodiphenyl-methane > 4, 4'-diaminodiphenyl ether > p-phenylene diamine > m-phenylene diamine > m-toluylene diamine > 4, 4'-diamino-3, 3'-dimethyldiphenyl-methane > 4, 4'-diaminodiphenyl sulfone.

It has been recently shown that aromatic polyamido-acids may be obtained in the absence of a solvent by application of high pressures /94/. It was shown that under a pressure of $6260\,kg/cm^2$ and at $200°$ pyromellitic dianhydride reacts with o- and m-phenylene diamines with the formation of polyamido-acids:

The resulting polymers are soluble in dimethylformamide. At $25°$ polymer I has an intrinsic viscosity of $[\eta] = 0.42$, while polymer II has $[\eta] = 1.02$. IR spectra confirmed that these polymers were polyamido-acids.

Conversion of polyamido-acids to polyimides. The process of conversion of polyamido-acids to polyimides (the second stage of the synthesis) is known as the imidization reaction (cyclodehydration) and consists of intramolecular evolution of water from the polyamido-acid with the formation of cyclic polyimide. The imidization reaction may be conducted in two ways: thermal and chemical.

The thermal method of imidization /10, 42, 101/ usually consists in heating the dried polyamido-acid, the rise in temperature being gradual or stepwise. Curing at high temperatures above $200°$ is conducted in vacuo or in an inert medium. Examples of typical experimental conditions employed for the purpose will be given in the following chapters. The imidization process may be controlled by following the changes in the IR or UV spectra of the polyamido-acid and also by following the evolution of water during the imidization /113/.

The chemical method of imidization /42, 60/ consists in treating the polyamido-acid powder or film with dehydrating agents. Such agents may be acetic anhydride or anhydrides of other lower fatty acids such as propionic, valeric, etc. Mixtures of different anhydrides may also be used, and also mixtures with aromatic monocarboxylic acid anhydrides. Ketenes and dimethylketenes may also be employed as dehydrating agents. Tertiary amines are employed as catalysts during chemical imidization. Pyridine, 4-methylpyridine, 3, 4-lutidine and isoquinoline are introduced in equimolecular ratio with the dehydrating agents. Amines such as 2-ethylpyridine, 2-methylpyridine, 2, 4-lutidine, and triethylamine are less reactive and larger amounts of them are introduced. Trimethylamine and triethylene diamine are very reactive and are employed in small amounts. The imidization process may be performed, for example, as follows /60/. The polyamido-acid film is treated for several hours at room temperature with a mixture of benzene, acetic anhydride and pyridine taken in equimolecular amounts. After the termination of the process the film is dried and heated at $300-500°$ for not less than one hour.

12

Chemical imidization may also be conducted in the solution of the polyamido-acid in dimethylformamide /66/, after addition of acetic anhydride and pyridine. After a certain time has elapsed, the polyimide gel separates out of solution; this is dried and thermally treated.

Individual aromatic polyimides obtained by the two-step method

Polypyromellitimides. By now numerous polyimides on the base of pyromellitic acid dianhydride — polypyromellitimides — have been prepared and studied.

The syntheses of polypyromellitimides from a number of aromatic diamines, as well as their properties, have been studied /42/. The polycondensation was effected in a solution of dimethylacetamide. The resulting polyamido-acids had a reduced viscosity between 1 and 3.0 and gave elastic films in most cases. Imidization of the polymers was effected both thermally and chemically. The main stress was laid on the study of the dependence of thermal stability of the polyimides thus obtained on the structure of the aromatic diamines and the nature of the grouping which interlink the phenyl nuclei. The different bonds and groups could be arranged in the following sequence of thermal stability in polypyromellitimides: phenyl-phenyl, imide > amide, ether > methylene > isopropylidene. It was also shown that the positions of the substituents in the benzene rings had no significant effect on the thermal stability of the polymers, but affected their elasticity. para-Derivatives of benzene typically form more viscous solutions.

Sroog et al. /101/ conducted a detailed study of polypyromellitimides both at the polyamido-acid and at the polyimide stage. They showed that the molecular weight of the polyamido-acid is determined by the experimental conditions of the reactions. The effect of temperature and time of reaction on the logarithmic viscosity of a polyamido-acid obtained from pyromellitic dianhydride and 4,4'-diaminodiphenyl ether in a dimethylacetamide solution is shown in Table 4.

TABLE 4. Effect of reaction conditions on the viscosity of polyamido-acid /101/

Concentration of solid, %	Temperature, °C	Time of reaction, min	η_{log}
10	25	120	4.05
10	65	30	3.47
10.6	85—88	30	2.44
10.7	115—119	15	1.16
10.3	125—128	15	1.00
15.7	135—137	15	0.59
14.2	150—160	2	0.51

It is seen from these data that an increase in the reaction temperature results in a considerable decrease in the logarithmic viscosity. This is in agreement with the data of /42/, obtained by a study of the polycondensation of pyromellitic dianhydride and 4,4'-diaminodiphenylmethane.

13

The polyimides prepared by Sroog et al. /101/ had an exceedingly high thermal stability. According to the data of thermogravimetric analysis, fully aromatic polyimides began to decompose only above 500°. The introduction into a diamine of a methylene group between the benzene rings reduced the incipient decompostion temperature to 450° (Figure 4). When heated for a long time (15 hours) polypyromellitimide obtained from 4,4'-diaminodiphenyl ether, lost only 1.5% of its own weight at 400°, 3% at 450° and 7% at 500°. Table 5 shows some of the physical parameters of polypyromellitimides obtained in /101/. Polypyromellitimides of different structures have also been prepared and studied by the authors of this book /19, 20, 30 and Chap. III/.

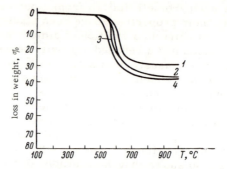

FIGURE 4. Effect of the structure of the diamine on the loss in weight of polypyromellitimides during heating at 3 deg/min
in helium /101/.

Most polypyromellitimides dissolve only in strong concentrated acids, and the dissolution is accompanied by degradation. Korshak et al. /8/ prepared polypyromellitimides soluble in organic solvents. These polymers were prepared from aromatic diamines — bis-(4-aminophenyl)-phthalide and bis-(4-aminophenyl)-phthalimide:

The synthesis of polyamido-acids is conducted in solution in N,N-dimethylacetamide or dimethyl sulfoxide. The imidization was effected thermally and chemically. The resulting polyimides were dissolved in dimethylformamide, dimethylacetamide and dimethyl sulfoxide even at room temperature and had the logarithmic viscosity of 0.4. The logarithmic viscosity of polyimides was lower than the viscosity of the corresponding polyamido-acids. This may be due to structural changes taking place during imidization and to the possible hydrolysis of polyamido-acids during the cyclization. Thermogravimetric analysis showed that the resulting polyimides do not lose weight even when heated at 450°.

TABLE 5. Properties of polypyromellitimides of the general formula $\left[-N\diagup\begin{smallmatrix}CO\\CO\end{smallmatrix}\diagdown\diagup\begin{smallmatrix}CO\\CO\end{smallmatrix}\diagdown N-R'-\right]_n$ according to /101/

R	Solvent	Mode of crystallization	Zero strength temperature,* °C	Thermal stability in the air	
				275°	300°
	H_2SO_4**	Crystallizes	900	>1 year	> 1 month
	H_2SO_4**	"	900	>1 year	–
	HNO_3	Readily crystallized	>900	–	1 month
$-CH_2-$	H_2SO_4	Crystallizes with difficulty	800	–	7–10 days
$-C(CH_3)_3-$	H_2SO_4	The same	580	–	15–20 days
$-S-$	HNO_3	Crystallizes	800	10–12 months	6 weeks
$-O-$	HNO_3	"	800	>1 year	> 1 month
$-SO_2-$	H_2SO_4	–	–	–	> 1 month
$-SO_2-$	H_2SO_4	–	–	–	> 1 month

* Defined as temperature at which the film withstands a load of 1.5 kg/cm^2 during 5 seconds.
** For the amorphous form (crystalline form is insoluble).

Polyarimides based on other tetracarboxylic acid dianhydrides. The use of the dianhydrides of different aromatic tetracarboxylic acids may serve to modify the properties of both the polyamido-acids and of the resulting polyimides.

Highly temperature-resistant polymers based on the dianhydride of 3, 3', 4, 4'-diphenyltetracarboxylic acid and aromatic diamines of the structure

$$\left[-N\diagup\begin{smallmatrix}CO\\CO\end{smallmatrix}\diagdown\diagup\begin{smallmatrix}CO\\CO\end{smallmatrix}\diagdown N-R'-\right]_n$$

were studied by Adrova et al. /1/. Polyamido-acids in dimethylformamide solution had intrinsic viscosities $[\eta]$ = 0.5 to 0.7. Imidization was performed by heating the polyamido-acid films to 350°, with a gradual rise in the temperature. Owing to the presence of a meta-bond between benzene rings in the dianhydride component, the polymer chain based on the dianhydride of 3, 3', 4, 4'-diphenyltetracarboxylic acid is more flexible than the chain based on the dianhydride of pyromellitic acid. In this way firm films may be produced from readily available aromatic diamines such as p-phenylene diamine and benzidine.

Synthesis of polyimides of the structure

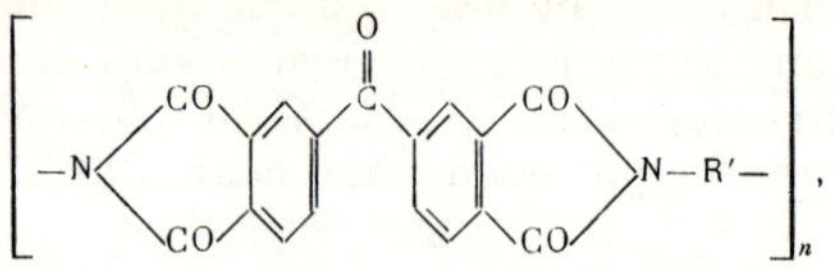

$$\left[-N \underset{CO}{\overset{CO}{<}} \underset{}{\bigcirc} \overset{\overset{O}{\underset{C}{||}}}{} \underset{}{\bigcirc} \underset{CO}{\overset{CO}{>}} N-R'- \right]_n ,$$

where

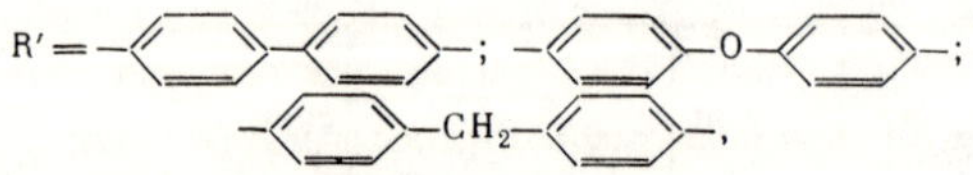

$$R' = -\bigcirc-\bigcirc- ; \quad -\bigcirc-O-\bigcirc- ;$$
$$-\bigcirc-CH_2-\bigcirc- ,$$

based on the dianhydride of 3,3',4,4'-benzophenonetetracarboxylic acid and aromatic diamines has been described /14, 56/. Benzidine-based polyimide does not soften. Diamines containing an oxygen atom and a methylene group* between the phenyl nuclei, give polyimides with a softening temperature of about 280°.

Polyimides based on 3,3', 4,4'-benzophenonetetracarboxylic acid and different aromatic diamines were also prepared and studied by the authors of the present book. These authors also prepared a number of polyimides based on the dianhydride of 3,3', 4,4'-diphenyloxidetetracarboxylic acid and aromatic diamines /4, 20, 30/. These polymers had the structure:

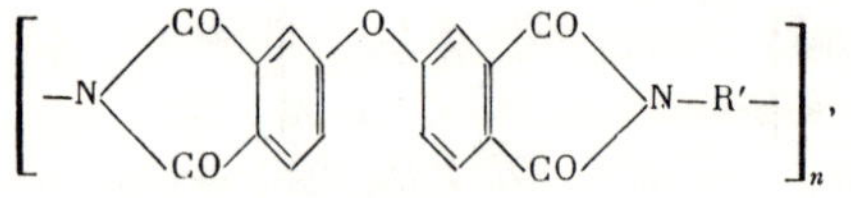

$$\left[-N \underset{CO}{\overset{CO}{<}} \underset{}{\bigcirc} O \underset{}{\bigcirc} \underset{CO}{\overset{CO}{>}} N-R'- \right]_n ,$$

where

$$R' = -\bigcirc- ; \quad -\bigcirc- ; \quad -\bigcirc-\bigcirc- ;$$
$$-\bigcirc-S-\bigcirc- ; \quad -\bigcirc-O-\bigcirc- ;$$
$$-\bigcirc-O-\bigcirc-O-\bigcirc- \quad \text{etc.}$$

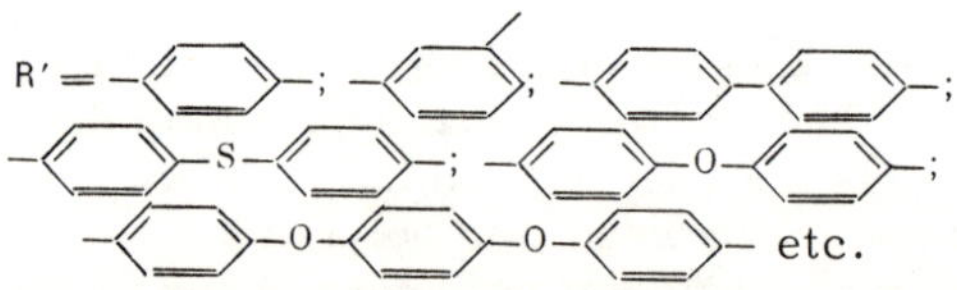

Owing to the presence of oxygen between the benzene rings in the dianhydride, polyimides based on the dianhydride of 3,3', 4,4'-diphenyloxidetetracarboxylic acid are more elastic and have lower softening temperatures than do polyimides obtained from pyromellitic and 3,3', 4,4'-diphenyltetracarboxylic acids. If the amine radical contains oxygen atoms between the benzene rings, the polyimides have a sharp softening point.

Kudryavtsev et al. /21/ prepared polyimides based on the dianhydride of 1,2,3,4-butanetetracarboxylic acid and aromatic diamines. The polyimides had the structure

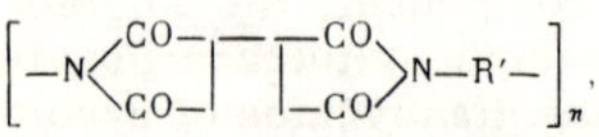

$$\left[-N \underset{CO-}{\overset{CO-}{<}} \underset{-CO}{\overset{-CO}{>}} N-R'- \right]_n ,$$

---

* ["Methyl" in Russian.]

16

where

The intrinsic viscosity of polyamido-acids was $0.3-0.5$ (in dimethyl-formamide at $20°$). The imidization was performed thermally, in the temperature range between 50 and $300°$. The polymers gave firm, elastic films.

Kolesnikov et al. /16/ prepared polyimides on the base of 2, 3, 5, 6-diphenyltetracarboxylic acid dianhydride and benzidine or 4, 4'-diamino-diphenylmethane. When the reaction was conducted in dimethylformamide at $30-100°$, polyamido-acids of a very low molecular weight were obtained ($\eta_{sp} < 0.074$), which may be due to the poor stability of the complex of this dianhydride with dimethylformamide. High-molecular polymers could be obtained in dimethyl sulfoxide — a solvent, which is more highly polar than dimethylformamide. The reaction temperature was $35-40°$. The structure of the polyimides was

The degree of polymerization and the molecular weight of the resulting polymers were determined from the acid and amine numbers obtained for polyamido-acids. The degree of polymerization n was calculated according to the formula:

$$n = \frac{[COOH] \cdot 16.02}{2\,[NH_2] \cdot 45.02},$$

where [COOH] is the content of COOH-groups in % and [NH$_2$] is the content of NH$_2$-groups in %.

The molecular weight of the polyamido-acid prepared from benzidine is 36,800; that of the polyamido-acid prepared from 4, 4'-diaminodiphenyl-methane was 10,500.

A patent /43/ describes the synthesis of aromatic polyimides with unreactive terminal groups. The synthesis was effected by the usual two-step method in polar solvents, in the presence of chain-terminating agents which are added to the reaction mixture in amount of $1-8$ mol.%. The following monoamines and acid anhydrides were used in the process: aniline, 4-aminophenyl ether, 4-aminodiphenyl, 4-aminodiphenylamine, 4-aminodiphenyl sulfide, phthalic anhydride etc. Polyimides thus obtained have the structure

In this way the molecular weight of polyimides can be controlled and cross-linking at the terminal groups prevented.

Plonka and Al'brekht /26/ prepared polyimides from 1, 4, 5, 8-naphthalene-tetracarboxylic acid dianhydride and aromatic diamines. Their structure may be represented by the formula:

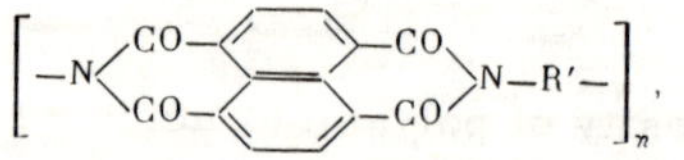

where

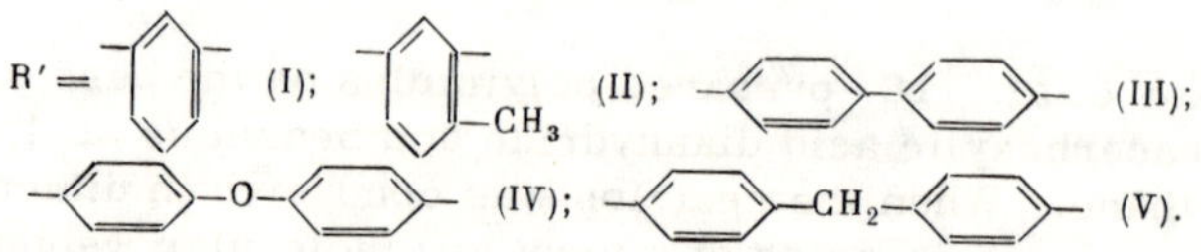

The resulting polymers are light-yellow (III and V) or gray-brown (I and II) powders, insoluble in the common organic solvents. Polyimides III and V yielded firm, transparent films.

Syntheses of a large number of polyarimides of different chemical structures are also described in the patents listed at the end of this book.

Polyimides with amido and ester groups in the chain. Recently attempts have been made to modify aromatic polyimide chains by introducing different links — mainly amide and ester links — into the main chain. The thermal stability of such polymers is intermediate between polyimides and polyamides (or polyesters). Polyamido-imides and polyester-imides, similarly to polyimides, have found extensive practical applications.

The following starting compounds /111/ may be employed in the preparation of polyester-imides:

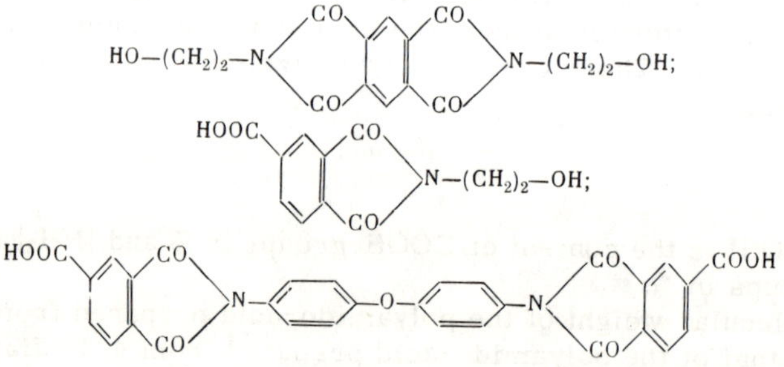

The reaction between the free carboxyl group of trimellitic dianhydride and diacetoxy compounds yielded /77/ new aroylene-bis-(trimellitate)-dianhydrides of the general formula

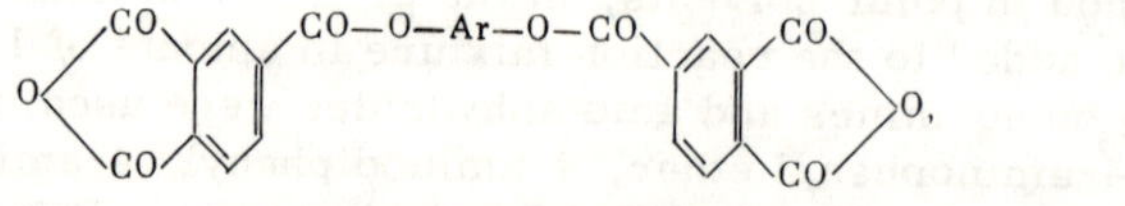

where

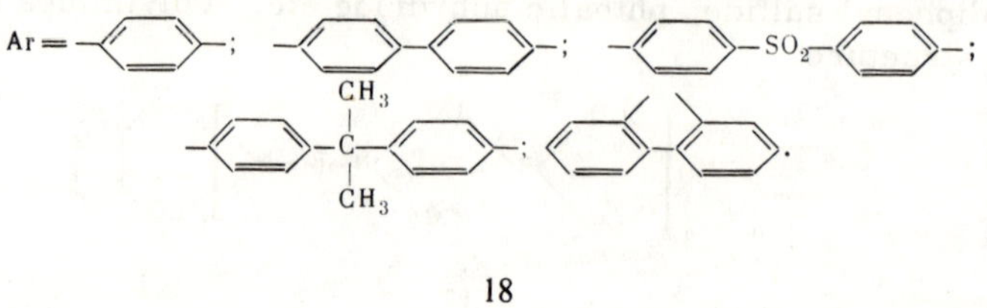

To an equimolecular mixture of diamine and dianhydride enough solvent was added to produce a 10—20% concentration of the solid matter. The polymers were prepared from p- and m-phenylene diamines, benzidine, 4,4'-diaminodiphenyl ether, 4,4'-diaminodiphenyl sulfone, 4,4'-diamino-diphenylmethane and other aromatic diamines. The resulting poly-steramido-acids had high logarithmic viscosities. Their thermal treatment gave polyester-imides with superior physical parameters and high thermal stability. Most resulting polymers form transparent elastic films. Better properties are displayed by films prepared from 4,4'-diaminodiphenyl ether. Polyester-imides are more thermostable than polyamido-imides.

Polymers with both amide and imide links in the chain have been obtained by the reaction between pyromellitic dianhydride and the dihydrazides of isophthalic /72, 105/, terephthalic /72/ and adipic /72/ acids according to the general scheme:

$$O \Big\langle {}^{CO}_{CO} \Big\rangle \Big\langle {}^{CO}_{CO} \Big\rangle O + H_2N-NH-CO-R'-CO-NH-NH_2 \longrightarrow$$

$$\longrightarrow \left[\begin{array}{c} -CO-NH-NH-CO \diagdown \diagup COOH \\ \\ HOOC \diagup \diagdown CO-NH-NH-CO-R'- \end{array} \right]_n \xrightarrow[-2nH_2O]{\Delta}$$

$$\longrightarrow \left[-CO-NH-N \Big\langle {}^{CO}_{CO} \Big\rangle \Big\langle {}^{CO}_{CO} \Big\rangle N-NH-CO-R'- \right]_n .$$

Dihydrazides of isophthalic and adipic acids react with pyromellitic anhydride to form polyamido-acids with a reduced viscosity of 0.5, in an almost quantitative yield. Terephthalyl dihydrazide failed to yield a high viscosity polymer. Dihydrazides with a low basicity are more difficult to react with pyromellitic dianhydride than are aromatic diamines of a similar structure. Thus, unlike isophthalyl dihydrazide, m-phenylene diamine forms a polyamido-acid with a viscosity of 3.2. The resulting polyamido-imides are not soluble in any solvent and swell in dimethyl-acetamide and m-cresol. Thermogravimetric investigations showed that aromatic polyamido-imides lose 5% of their own weight up to 380°. Between 380 and 560° a large decrease in weight is noted. The aliphatic polymers begin to show rapid decomposition at 360°.

Polycondenstaion of phenylene-bis-(trimellitate)-dianhydride with isophthalic acid dihydriazide yielded /78/ a polyimide containing both amide and ester groups in the chain:

$$NH_2-NH-CO \diagdown \diagup CO-NH-NH_2 +$$

$$+ O \Big\langle {}^{CO}_{CO} \Big\rangle CO-O-\langle \hspace{0.5em} \rangle -O-CO \diagup \Big\langle {}^{CO}_{CO} \Big\rangle O \longrightarrow$$

$$\longrightarrow \left[\begin{array}{c} CO-NH-NH-CO \diagdown \diagup CO-O-\langle \hspace{0.5em} \rangle -O-CO \diagdown \diagup COOH \\ \\ HOOC \diagup \hspace{2em} CO-NH-NH-CO- \end{array} \right]_n \xrightarrow{\Delta}$$

$$\xrightarrow{\Delta} \left[\begin{array}{c} CO-NH-N \diagdown \diagup CO-O-\langle \hspace{0.5em} \rangle -O-CO \diagdown \diagup \\ {}^{CO}_{CO} \hspace{6em} {}^{CO}_{CO} \end{array} N-NH-CO- \right]_n .$$

When the reaction is conducted in dimethylformamide, a viscous solution of polyester-hydrazido-acid is formed. Transparent films may be formed from this solution. The polyester-hydrazido-acid is also soluble in dimethylacetamide, dimethyl sulfoxide, N-methyl-2-pyrrolidone and pyridine. The conversion of the polyester-hydrazido-acid to the polyester-amido-imide is effected by the thermal technique. Thermogravimetric studies showed that the decomposition of the polymer in the air begins at 350° (Figure 5).

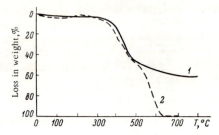

FIGURE 5. Thermogravimetric curves of polyesteramido-imide in nitrogen (1) and in the air (2) /78/.

This group of polyimides is related to polymers with urea groups in the chain /109/. They are prepared by reacting pyromellitic acid dianhydride with carbonic acid dihydrazide, as follows:

$$O \underset{CO}{\overset{CO}{\diagdown}} \underset{CO}{\overset{CO}{\diagup}} O + H_2N-NH-CO-NH-NH_2 \longrightarrow$$

$$\longrightarrow \left[\begin{array}{c} -NH-CO \diagdown \\ HOOC \diagup \end{array} \underset{CO-NH-NH--CO-NH-}{\overset{COOH}{\diagup}} \right]_n \overset{\Delta}{\longrightarrow}$$

$$\longrightarrow \left[-N \underset{CO}{\overset{CO}{\diagdown}} \underset{CO}{\overset{CO}{\diagup}} N-NH-CO-NH- \right]_n .$$

Polypyromellitimino-urea obtained in a solution of N,N-dimethylacetamide or dimethyl sulfoxide had an intrinsic viscosity of 0.22—0.47 (in 1N NaOH at 30°) and gave a flexible, nonfusible film. The conversion of polypyromellitimino-urea to polypyromellitimido-urea was performed by the thermal method. At 280—310° an oxidative degradation of the polymer was noted; above 380° the polymer is fully decomposed.

Preparation of polyimides from dihalides and aromatic diimides

Nishizaki and Fukami /88, 89/ prepared polyimides by reacting pyromellitic acid diimides with different dihalides. The polycondensation is accompanied by the separation of hydrogen chloride (or hydrogen bromide) and proceeds as follows:

where

The reaction is catalyzed by tertiary amines (e.g., triethylamine) and potassium carbonate. It is conducted in a solution of dimethylformamide, with heating. If ethylene dichloride is employed, the yield of the reaction is 80%. Aromatic dichlorides do not enter the reaction even after prolonged heating. Aromatic chloromethyl derivatives react just as readily as the aliphatic dihalides. The resulting polyimides were insoluble and their molecular weight was low.

Reaction between bis-halomethyldisiloxanes

and pyromellitic acid diimide yielded /88/ silicon-containing polyimides in good yields. Table 6 gives the properties of silicon-containing polyimides obtained from pyromellitic acid diimide and chloromethyl-disiloxane under various conditions. Similar results were obtained for the reaction between pyromellitic acid diimide and bromomethyldisiloxane.

TABLE 6. Properties of silicon-containing polypyromellitic acid imides /88/

Catalyst	Reaction conditions	Yield, %	T_f, °C	Reduced viscosity
K_2CO_3	130°, 5 hours	47	115	0.10
$(C_2H_5)_3N$	150°, 5 "	65	120	0.18
$(C_4H_{10})_3N$	150°, 5 "	56	115	0.15
Pyridine	150°, 5 "	50	115	0.10

Note. Melting points determined by the capillary method.

Silicon-containing polypyromellitic acid imides are soluble in dimethylacetamide and dimethyl sulfoxide and insoluble in methanol and ether. The catalysts employed in the polycondensation reaction can be arranged in the following decreasing sequence of activity: triethylamine > tributyl-amine > pyridine > K_2CO_3.

Polyimides have also been prepared /88, 89/ by the reaction between potassium pyromellitic acid diimide and p, p'-dichloromethyldiphenyl ether:

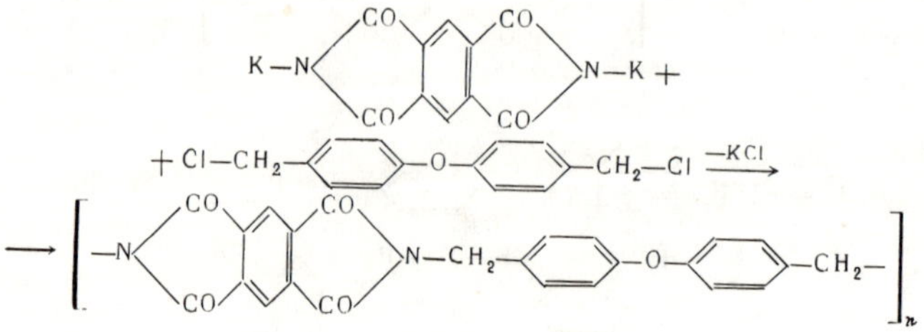

The reaction was rapid and gave high yields of the polymer at high temperatures. The polymer yield after 8 hours was 85.4% at 140°, 80% at 100° and 74% at 70°.

Some new nitrogen-containing polymers based on amines and dianhydrides

The preparation and properties of a new class of heat-resistant and radiation-resistant polymers of a stepladder structure — polyimidazo-pyrrolones (pyrones) — are similar to those of the polyimides. These polymers are prepared by the reaction between aromatic tetraamines and dianhydrides of aromatic tetracarboxylic acids. Depending on the structure of the initial compounds, the polymers may have a fully stepladder struc-ture or consist of stepladder segments of not less than four rings, connected by different links such as —CO—, —O— etc.

The formation of polymers of this class may be illustrated by the reaction between pyromellitic acid dianhydride and 3, 3'-diaminobenzidine as follows:

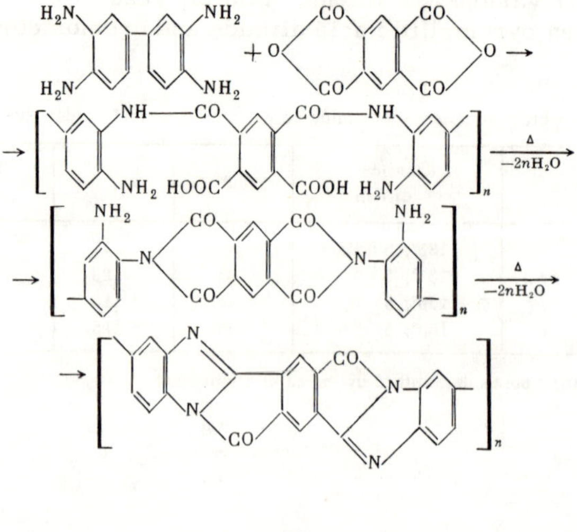

22

Most studies in this field /39, 53, 57, 59/ deal with the synthesis of stepladder polymers from pyromellitic acid dianhydrides and benzophenone-tetracarboxylic acid and aromatic tetraamines: 3, 3'-diaminobenzidine, 1, 2, 4, 5-tetraaminobenzene, 3, 3', 4, 4'-tetraaminodiphenyl ether and 1, 4, 5, 8-tetraaminonaphthalene.

The chief method employed in the synthesis of pyrones is the three-step polycondensation method. In a solution of dimethylacetamide, or another polar solvent, the soluble polyamido-acid is obtained (logarithmic viscosity $1-1.5$), from which films can be prepared. Ring closure takes place when the film is heated. At $130-150°$ polyamino-imide is formed and at $225-250°$ the polyimidazopyrrolone is obtained. The resulting polymers are insoluble in the conventional organic solvents, but are soluble in dimethyl sulfoxide and in sulfuric acid.

Pyrones were also obtained by polycondensation in polyphosphoric acid and in the melt /59/.

Pyrones are stable when heated to 500° in the air. As an example, Figure 6 gives thermogravimetric curves of pyrones prepared under various conditions /59/. Pyrones are remarkably resistant to radiation; they withstand radiation doses of up to 10^{10} rad.

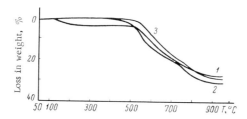

FIGURE 6. Thermogravimetric curves of pyrones /59/.

Polymers obtained in the melt (1, 2) and in 116% polyphosphoric acid (3).

Berlin et al. /5/ and Vandusen /110/ polycondensed the dianhydride of 1, 4, 5, 8-naphthalenetetracarboxylic acid with 1, 2, 4, 5-tetraaminodiphenyl to pyrones of the structure:

These polymers are black and are soluble in sulfuric and polyphosphoric acids. They are stable on being heated to 400—500° in the air and up to 600° on being heated in nitrogen.

Preston and Black /93/ prepared a new class of linear heterocyclic thermostable polymers, in which the imide rings in the polymer chain regularly alternate with other heterocyclic units, e.g., oxadiazole — benzimidazole, oxadiazole — pyromellitic acid imide, thiazole — pyromellitic acid imide. These polymers were obtained by reacting pyromellitic acid dianhydride with aromatic diamines containing hetero-cyclic links. The structure of the resulting polymers may be represented by the following formulas:

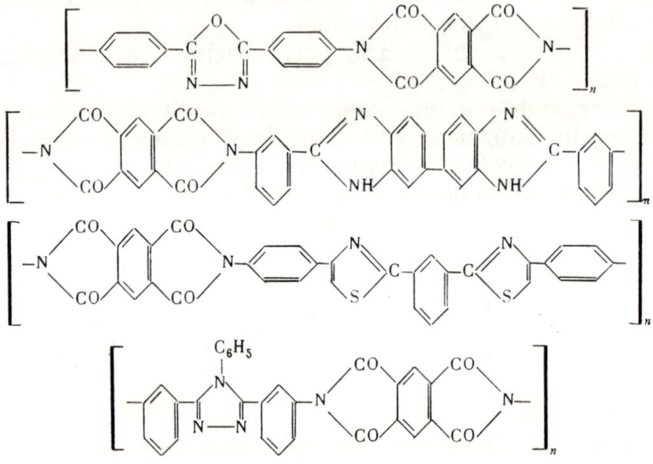

These polymers do not melt and do not decompose below 500°.

Imidization kinetics and secondary chemical transformations of polyimides

It has been shown above that the second step of the synthesis of polyimides (imidization, cyclodehydration) is performed by thermal or chemical treatment of the polyamido-acid. This is accompanied by the evolution of water and formation of polyimide:

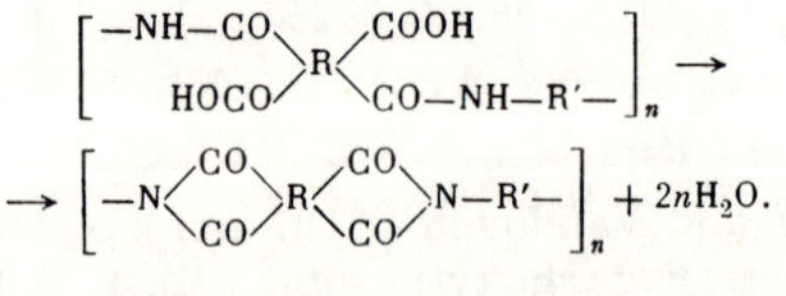

Imidization may take place both in solution and in the solid phase. In the latter case the polyamido-acid must first be isolated from solution. In practice, polyimides are mostly obtained by imidization in the solid phase. A discussion of the results obtained in the study of the mechanism and kinetics of the imidization reaction and a number of data on the subsequent chemical modifications of polyimides are given below.

Wrasidlo et al. /113/ studied the kinetics of formation of polyamido-acid from pyromellitic dianhydride and m-phenylene diamine in a solution of dimethylacetamide and also the imidization kinetics of this acid in the same solvent. The kinetics of formation of the polyamido-acid were studied at room temperature by following the temporal variation in the concentration of the dianhydride, as indicated by the characteristic IR-band at $1836\,cm^{-1}$. The degree of imidization was determined at $80°$ by determining the amount of water evolved by vapor phase chromatography, and also qualitatively by following the variation of the intensity of the $1780\,cm^{-1}$ absorption band.

The authors showed that in both cases the usual kinetic equation of the type

$$v = k \cdot C^n,$$

applies, where v is the transformation rate, C is the concentration of the reacting substance, k is the rate constant and n is the order of the reaction.

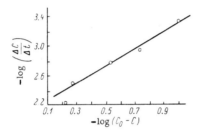

FIGURE 7. Concentration dependence of the imidization rate of a polyimide based on pyromellitic acid dianhydride and m-phenylene diamine in dimethylacetamide /113/.

c — concentration of imidization water evolved at the moment of the determination; c_0 — water concentration corresponding to complete imidization.

It was found that the formation of the polyamido-acid is a second order reaction $(n \simeq 2.16)$ and that the rate constant of the reaction is $k = = 1.06$ liters/mole·sec at room temperature. Imidization in solution is an apparent first order reaction $(n \simeq 1.1)$ with a rate constant of $3.87 \cdot 10^{-3}\,sec^{-1}$ at $80°$, as can be seen from Figure 7. The reactions were very vigorous. Thus, for example, the half-conversion period during imidization at $80°$ was $\tau = \frac{\ln 2}{k} = 34\,sec$; within 11 minutes the experimental degree of imidization attained 98.5%. In both cases the experimental values of the order of the reaction are in full agreement with the commonly assumed reaction schemes representing the conversions resulting from amidization and imidization. The authors /113/ assumed, however, that the main imidization reaction is accompanied by a side reaction, which consists in the formation of an intermediate salt-type compound (II) which can be detected by its characteristic NH_3^+-grouping absorption bands at 2530 and 2650 cm^{-1}. This compound evolves water with the formation of a branched polyamide chain (III) which in turn yields the linear polyimide chain (V):

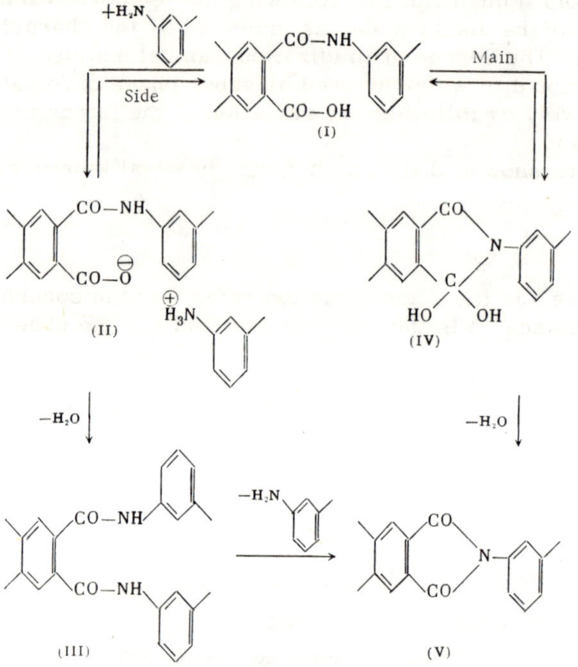

The authors /113/ do not bring forward any experimental evidence in support of the existence of the two last-named reactions or of the intermediate compound (IV).

At room temperature imidization is very slow. Thus, in a solution of polyamido-acid prepared from pyromellitic acid dianhydride and m-phenylene diamine no imidization was noted after 12 hours' standing at 25° /113/. Polyamido-acid from pyromellitic acid dianhydride and 4,4'-diaminodiphenyl ether /66/ was imidized to the extent of only 20% after 5000 hours at 35°.

The water formed in solution during imidization is not removed from the reaction mixture and may bring about hydrolytic cleavage of macromolecules and decrease of the molecular weight of the polymer at high temperatures /66, 87/. This consideration must be borne in mind when effecting imidization in the solid phase, since the working temperature is then high — up to 250—300°. In order to prevent hydrolysis, the water formed must be immediately removed from the system. To do this, imidization of solid polyamido-acids should be performed in thin layers, i.e., polyamido-acids should be used as fibers, powder or films not more than 100—200μ thick.

A study of the thermal imidization kinetics in the solid phase based on the IR absorption spectra of the films was performed by the authors of this book on polyimide PM obtained from pyromellitic dianhydride and 4,4'-diaminodiphenyl ether /23/. This study will be discussed in detail below, since despite the reports /64, 100 etc./ on the applicability of IR spectroscopy to the study of the imidization process, quantitative data on the subject in the literature are scanty.*

Polyamido-acid films were cast from solution and dried at 80° during 10 minutes. The film thickness was $2-2.5\mu$.

Imidization was carried out in a stream of nitrogen in a special thermal chamber. The time of the thermal treatment was determined to within $\pm 2-3$ seconds, and the temperature to within $\pm 0.5°$. The spectra were recorded at room temperature with the aid of an IKS-14 instrument. Film samples were preliminarily heated for 5 minutes at 160° to remove the solvent and to bring them to a standard initial state.

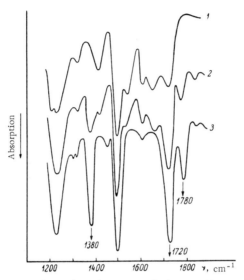

FIGURE 8. IR absorption spectra of polypyromellit-imide prepared from pyromellitic dianhydride and diaminodiphenyl ether at different stages of imidiza-tion.

1 — initial film (poly-acid); 2 — film heated at 187°C for 4 minutes; 3 — film heated at 300°C for 15 minutes (polyimide).

The IR spectra of PM polyimide (Figure 8) contain absorption bands typical of the imide ring: $1380\,\mathrm{cm}^{-1}$ and $1780\,\mathrm{cm}^{-1}$ (doublet with $1720\,\mathrm{cm}^{-1}$), which correspond to vibrations of cyclic $C-N$ and $C=O$ respectively /3,86,87/. The course of the imidization was followed by recording the intensities of these bands.** The band intensities no longer increased after the initial sample had been heated at 300° for 15 minutes. This was assumed to be the final standard state.

---

* For another study of solid phase imidization kinetics by IR spectroscopy, see (J.Polym. Sc., A-1, 4(9): 2609. 1966). The conclusions on the typical features of solid phase imidization arrived at in this study are in agreement with those of /23/. In another recent study (Vysokomolekulyamye Soedineniya, 9B(3):201. 1967) the kinetics was studied by following the variation in the film weight.

** The intensities were measured from the peak of the absorption band to the "zero" base line marked by the initial contour of the band corresponding to the polyamido-acid.

The degree of imidization i was defined as

$$i = \frac{D}{D_{300}},$$

where D is the optical density on the characteristic band of the sample heat-treated for a given time at a given temperature, and D_{300} is the optical density of the fully imidized sample (heated at 300° for 15 minutes).

For polyamido-acid $i = 0$, for the polyimide $i = 1$. For films in the initial standard state $i \simeq 0.10$.

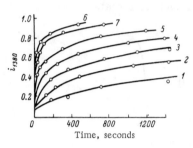

FIGURE 9. Variation of the degree of imidization of the polyimide PM, calculated from the $1380\,cm^{-1}$ band, with time during the thermal treatment.

1 — 162°C; 2 — 187°C; 3 — 200°C;
4 — 212°C; 5 — 230°C; 6 — 250°C;
7 — 230°C (without preliminary heating).

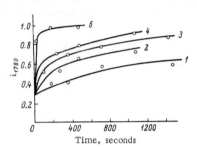

FIGURE 10. Degree of imidization of polyimide PM, calculated from the $1780\,cm^{-1}$ band, on the time of thermal treatment.

Numbering of curves is the same as in Figure 9.

Figures 9 and 10 indicate the variation of the degree of imidization i during the thermal treatment of the films at different temperatures. As the duration of the heat treatment increases, the imidization steadily increases; as the experimental temperature increases, the process becomes more rapid. It is noteworthy that the degrees of imidization calculated from the $1780\,cm^{-1}$ band are always higher than those calculated from the $1380\,cm^{-1}$ band. The differences are especially large at the initial stages of imidization. This can also be seen from Figure 11, which gives a compariosn between the values of the degrees of imidization calculated from both bands for a large number of experiments conducted at various temperatures. Had both calculations given the same result, the experimental points would have been found on the straight line at a 45° angle (dotted line in Figure 11), i.e., we would have $i_{1780} = i_{1380}$. In fact, the result is invariably $i_{1780} > i_{1380}$, except for the reference points $i = 0$ and $i = 1$.

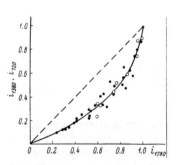

FIGURE 11. Comparison of the degree of imidization of the polyimide PM, calculated from 1380, 720 and $1780\,cm^{-1}$ bands.

$\bullet$ — $1380\,cm^{-1}$; $\circ$ — $720\,cm^{-1}$.

28

The nonsymbatic variation of the values of i_{1780} and i_{1380} is apparently due to Beer's Law not being obeyed by one of the bands in the process of imidization. It would appear that it is the absorption coefficient at the frequency of $1780\,\mathrm{cm}^{-1}$ (K_{1780}) which is variable in this case. It is known /25/ that the absorption coefficient of a carbonyl group increases in the presence of a hydrogen bond. In the early and intermediate imidization stages hydrogen bonds are quite liable to be formed between the oxygen atom of the carbonyl group of the imide ring and a hydrogen atom of the amide or hydroxyl group of polyamido-acid molecules:

For this reason, at low degrees of imidization the absorption coefficient K_{1780} and the degree of imidization i_{1780} may prove to be high. As the content of the polyamido-acid decreases, two processes will take place. The intensity of the absorption band will increase on one hand due to the forma- tion of new rings, and will decrease, on the other, owing to the decrease in the absorption coefficient. The superposition of these processes will result in the value of i_{1780} tending to unity long before the imidization process is actually completed. All this is in agreement with the data in Figure 11. If it is assumed that the absorption coefficient K_{1780} varies linearly with the degree of imidization, while the latter can be determined without error from the $1380\,\mathrm{cm}^{-1}$ band, we shall find from the data in Figure 11 that the coefficient K_{1780} decreases by approximately one-half on passing from the polyamido-acid to the polyimide. The effect of the hydrogen bond on the absorption coefficient of the $>\!C\!=\!O$ band in acetone is similar /25/.

Infra-red spectra of polyimides have another characteristic band at $720\,\mathrm{cm}^{-1}$ /64/. Its origin is not yet quite clear. The band is absent in the spectrum of the polyamido-acid and steadily increases in intensity as the thermal treatment proceeds. The imidization degree i_{720}, calculated from this band, is practically identical with i_{1380} (Figure 11).

These results show that the selection of the characteristic band is very important in quantitative measurements of the degree of imidization from the IR spectra. Apparently, it is preferable to use the $1380\,\mathrm{cm}^{-1}$ band ($C\!-\!N$ vibration) which is not affected by hydrogen bonds rather than the $1780\,\mathrm{cm}^{-1}$ band which is more often mentioned in the literature. A final answer can, however, be obtained only by comparing the spectroscopic data with the results of an independent absolute method of determination of the degree of imidization. In what follows the value of i_{1380} is employed as an index of the degree of imidization in determining the imidization kinetics. Calculation of kinetic characteristics from the values of i_{1780} gives qualitatively similar results.

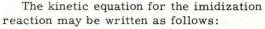

FIGURE 12. Dependence of the logarithm of imidization rate log v on the logarithm of incompleteness of imidization log $(1-i)$ at different temperatures for the reaction in the solid phase (PM polyimide).

The kinetic equation for the imidization reaction may be written as follows:

$$v = \frac{di}{dt} = k\,(1 - i)^n.$$

If during the ring formation the reacting COOH- and NH-groups are located in the same macromolecule, it is to be expected that the imidization reaction will be of the first order $(n = 1)$. In dilute solutions this is in fact very nearly the case /113/. If the imidization takes place in the solid phase, the problem of the order of the reaction becomes more difficult. In this case the function log v vs. log $(1-i)$ is curvilinear at all temperatures (Figure 12). The values of the index n, calculated from the average slope of the curve are quite different from unity and vary with the temperature between 2.2 and 3.2. If the imidization is conducted in solution, these relationships are linear, with a slope close to unity (cf. Figure 7).

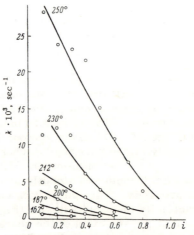

FIGURE 13. Imidization rate constant as a function of the degree of imidization at different temperatures (PM polyimide).

The reason for this effect is probably that the rate constant k of solid phase imidization does not remain constant at constant temperature as the reaction proceeds, i.e., is a function of the degree of imidization. If this is in fact the case, the function log v vs. log $(1-i)$ will be clearly nonlinear and cannot serve, in principle, for the calculation of the reaction order. The values of the rate constants at different degrees of conversion may be found from Figure 12 if it is assumed that the order of the imidization reaction is constant and equal to unity, as is the case in solution. It is seen from Figure 13 that the reaction constants monotonously decrease with increasing degree of imidization, at all experimental temperatures. The value of the constant decreases by a factor of $10-12$ as i varies between 0.2 and 0.8.

The decrease in the imidization rate constant with increasing i may be explained by assuming that the energetic and temporal parameters of the constant k in Arrhenius' equation

$$k = A e^{-\frac{U}{RT}}$$

do not remain constant in the solid phase. This can be seen by studying the temperature dependence of the imidization rate at different constant values of the degree of imidization (Figure 14). The function log v vs. $\frac{1}{T}$ is a straight line, i.e., Arrhenius' equation is valid, but the slope of the lines increases with increasing i.

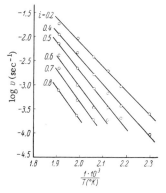

FIGURE 14. Logarithm of imidization rate as a function of the reciprocal temperature at different values of the degree of imidization (PM polyimide).

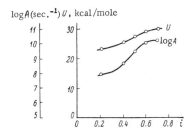

FIGURE 15. Activation energy U and logarithm of pre-exponential factor A as a function of the degree of imidization for the polyimide PM.

The activation energy U increases from 23 to 30 kcal/mole as i varies from 0.2 to 0.8, and is independent of the order of the reaction. The pre-exponential factor A increases at the same time. The variation of U and log A is represented in Figure 15, in which log A has been calculated on the assumption that $n = 1$.

The variation of the activation energy and the temporal factor may be due to several causes. It is possible that the ⟨CO—OH / CO—NH—⟩ groupings in the solid polyamido-acid have different but constant values of the energy

31

barrier and vibration frequency, for steric or some other reasons; at the beginning of the reaction (at a constant temperature) the most reactive groups react, while at later stages the least active groups are involved. As a result, the rate constant will gradually decrease. Another reason for this may be the fact that these parameters of the ⟨benzene ring with CO—OH and CO—NH— groupings⟩ groupings vary in the course of the reaction, perhaps owing to a decrease in the mobility of the macromolecules with the increase in their content of imide rings. The imidization rate constant will also decrease in this case. Finally, the rate of imidization in the solid phase depends on the presence of traces of solvent in the initial films of the polyamido-acid. It is seen in Figure 9 that in a sample which has not been pre-heated at 160° (curve 7) the conversion rate in the initial stages is much higher than in a sample which has been subjected to a preliminary thermal treatment (curve 5). If the only effect of the solvent is plasticization, this effect should be very small if the films are thin and the temperatures are high (in analogy with the effect of the water of imidization, see below). If the solvent is chemically bound to the polyamido-acid and participates directly in the imidization, e.g., through complex formation with COOH- and NH-groups which facilitates ring formation, the imidization rate will be affected even if the solvent is present in residual trace concentrations only. The rate constant will then also decrease with increasing degree of imidization (decrease in the content of the solvent). It is as yet difficult to say which of these factors actually affects the imidization in the solid phase.

There is yet another factor which must be considered. If the water evolved during the imidization is not sufficiently rapidly removed from the reaction zone, the imide may be reconverted to the amido-acid /66/. Then the overall rate constant of the imidization will be determined by the rates of the forward and the reverse reactions and by the rate of removal of the water; if the water removal is slow, it will be the rate-determining factor.

It can be shown that when thin films are employed, the effect of the reverse reaction is insignificant. The typical time τ of the removal of the liquid from the film is given by the expression /11/

$$\tau = a\,\frac{d^2}{D},$$

where d is the film thickness, D is the coefficient of diffusion and a is a constant factor of the order of 0.1—0.2.

For the system water—polymer, D is about 10^{-8} cm²/sec at 20° /11/, and if $d = 2\,\mu$, $\tau \sim 1$ sec. When the temperature is increased, D increases by about 1.5—2 order for each 100° so that τ should be of the order of 0.1—0.01 sec. Under these conditions the typical duration of time required for the conversion of the polyacid to polyimide (reciprocal of imidization rate) is 1000—50 sec. It follows that in thin films the course of the imidization process cannot be limited by the rate of water removal. It will be clear, however, that diffusional processes must be allowed for if the layers are thick.

It has been experimentally shown /23/ that the imidization mechanism in the solid phase is more complex than in solution. This is why the only magnitudes which could be accurately determined in this case are the activation energy and its variation in the course of the process. Other fundamental kinetic parameters could only be arrived at by making additional assumptions based on the results obtained with solutions, a procedure which is not unobjectionable.

Kudryavtsev et al. /21/ studied the thermal imidization of solid polyamido-acids based on the dianhydride of the (aliphatic) 1,2,3,4-butanetetracarboxylic acid by IR spectroscopy. As in the case of polypyromellitimide, the spectra underwent marked changes in the course of the thermal treatment. The course of imidization was followed (Figure 16) by following the change in the intensity of the absorption bands at $1780\,cm^{-1}$ (cyclic C = O vibration) $1645\,cm^{-1}$ (acyclic C = O vibration) and $1530\,cm^{-1}$ (NH-group vibration). Figure 17 represents data on the variation of the optical density of $1780\,cm^{-1}$ band relative to the optical density of the $1020\,cm^{-1}$ (reference) band observed during the thermal treatment of polymers prepared from this dianhydride and different diamines. In all cases the ratio $\frac{D_{1780}}{D_{1020}}$ increased (increase in the number of rings) and then decreased starting from certain temperatures, which indicates a decrease in the number of imide rings. Ring formation proceeds at the most intense rate in the case of polyamido-acids based on p- and m-phenylene diamines, and at a much less intense rate in the case of polyamido-acids based on benzidine and 4,4'-diaminodiphenyl ether.

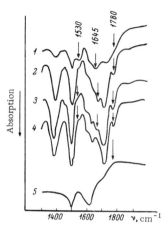

FIGURE 16. IR absorption spectra of the polyamido-acid prepared from the dianhydride of 1,2,3,4-butane-tetracarboxylic acid and 4,4'-diaminodiphenyl ether, at different stages of thermal treatment /21/.

1 — 25°C; 2 — 150°C; 3 — 200°C; 4 — 280°C; 5 — 350°C. Duration of heating: half an hour at each temperature.

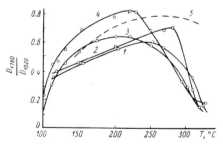

FIGURE 17. Variation of relative optical density during the thermal treatment of polyamido-acids based on the dianhydride of 1,2,2,4-butanetetra-carboxylic acid and different diamines: benzidine (1), 4,4'-diaminodiphenyl ether (2), m-phenylene diamine (3), p-phenylene diamine (4) /21/.

The dotted line represents data for polypyromellit-amido-acid based on 4,4'-diaminodiphenyl ether.

On the other hand, ring stabilities in the last-mentioned two cases proved to be much higher: the maxima of the curves $\frac{D_{1780}}{D_{1020}}$ vs. T shift by 50—100° towards higher temperatures. If an aromatic rather than aliphatic dianhydride is used, the imidization rate increases markedly and the ring stability proves higher (cf. curves 2 and 5, Figure 17).

It may be concluded from /21/ that the reactivity of $\underset{}{>}R\underset{CO-NH-R'-}{\overset{CO-OH}{<}}$

groups as far as imide ring formation is concerned depends on the structure of radicals R and R'. Ring formation is facilitated if the chian contains conjugated bonds and becomes more difficult if the degree of conjugation decreases.

It will be shown in Chapter III that the process of imidization is accompanied by marked changes in numerous physical properties (mechanical parameters, density, refractive index, dielectric parameters etc.). These changes may also be employed to study the relationships governing the imidization. It may be concluded, for example, from the results of dielectric and refractometric measurements (see below, Figures 41, 90, 91) that the rate of imidization in the solid phase and its temperature range are about the same for all aromatic polyimides.

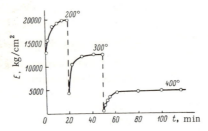

FIGURE 18. Temporal variation of the static elasticity modulus of the polypyromellit-imide

in argon at various temperatures /31/.

The chemical conversions in the polyimides are not limited to the imidization process. This process is practically completed at 250—300° within 10—15 minutes, at least for aromatic polyimides, as indicated by spectroscopic and other data. Nevertheless, the mechanical properties of films — strength, elasticity modulus and especially elasticity at low temperatures — often continue to show marked variations after heating at higher temperatures /31/. This subject will be discussed in detail in connection with the physical properties of polyimides. We shall merely show one graph (Figure 18), which illustrates the variation in the elasticity modulus of a sample of polymer based on pyromellitic dianhydride and 4,4'-diaminodiphenyl ether in the course of stepwise thermal treatment.

The increase in the elasticity modulus at 200 and 300°, followed by its sharp drop, may be explained by the occurrence of the imidization process, which results in an increased rigidity of the polymer chains. The considerable changes in the modulus at 400° cannot be connected with imidization, since the polymer may be considered as fully cyclized after heating at 300° This effect is also difficult to explain by purely physical changes, such as crystallization, since this polyarimide crystallizes only to an insignificant extent or not at all. Moreover, crystallization as a rule results in a decrease in elasticity, whereas in our case thermal treatment at 400° results in a marked increase in elasticity, especially at below-zero temperatures, while in the oriented state the material does not contract on being heated /31/. Cooper et al. /58/ also pointed out chemical changes in polyimides indicated by the changes in elasticity modulus.

It has been suggested by the authors of this book /31/ that these effects are connected with the formation of a special kind of structural network in polyimides by way of inter-chain isomeric transformations. It is assumed that at high temperatures the N—CO bonds in the imide rings are in a state of continuous disruption and reformation. The latter may result not only in the formation of intramolecular rings, but also in the formation of imide links between one chain and another, e.g.:

The formation of such network explains the increase of the elasticity modulus after the completion of imidization, loss of softening capacity and sag tendency of oriented polypyromellitimide films and is not in contradiction with the increased elasticity effect if it is assumed that the system of successive inter-chain N—CO bonds is sufficiently extended.

We may compare this scheme with certain experimental data concerning structure. For example, the decrease in the $1780\,cm^{-1}$ band intensity in the IR specrta of polyimides heated at 300—400° (Figure 17 and /31/) may be explained by decreased concentration of imide rings, as may be seen from the scheme. It was shown by Boldyrev et al. /6/ that heating the cyclized polyimide above 300° results in an increased concentration of stable radicals, which may be related to the nitrogen atom. It may also be related to the chemical reaction mechanism represented in the scheme. However, the formation of intermolecular N—CO bonds should be accompanied not only by a weakening of the typical imide bonds, but by the formation of new bands corresponding to noncyclic $C = O$ vibrations. This can be seen, for example, in the IR spectra (Figure 19) of the compound

35

which is a model of type (I) inter-chain bonds, such as may be formed after decomposition of imide rings in the aromatic polyimide. The spectrum of the model compound does not contain the bands at 1780 and 1380 cm^{-1} typical of the imide ring (cf. Figure 8) and contains an intense band at 1660 cm^{-1}. However, the appearance of this band in the spectrum of polypyromellitimide on intense heating was not noted.

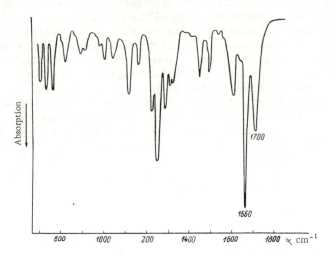

FIGURE 19. IR absorption spectrum of the model compound

Thus, the available structural data speak both for and against the scheme which has been given above for high-temperature secondary chemical transformations in polyimides, leading to the formation of inter-chain bonds. It would appear that no final conclusions can as yet be drawn as to the chemical mechanism of the cross-linking of polyimides at high temperatures, even though such cross-linking is undoubtedly present. A number of typical features of this process as reflected in the changes in the physical and mechanical properties of polyimides of different structures, will be described in Chapter III.

Chapter II

THERMAL AND CHEMICAL STABILITY OF
POLYIMIDES

   The high thermal stability of polyimides is their main advantage over
other high polymers and the primary reason for the interest they have
evoked. It is very important to understand the reasons for this stability;
since in this way syntheses of new polyimides can be correctly performed,
their maximum service potentialities determined, and many other practical
objectives achieved. Studies of the course and mechanism of decomposition
of polyimides at high temperatures — including both thermal degradation
and pyrolysis — are the most extensive and the most systematic which have
been published on the subject. These results are partly available in the
form of reviews /9, 27, 28/, and partly as uncritical nonsystematized
presentations.
   The first part of this chapter gives a detailed discussion of the most
important results obtained in the studies of the chemical mechanism of
thermal and thermooxidative degradation and pyrolysis of polypyromellitimides. The second part summarizes the data relating to the connection
between the chemical structure of polyimides and their thermal stability.
   The end of the chapter deals with the studies on the hydrolysis of
polyimides and with literature data on their chemical stability.

**Thermal degradation of polypyromellitimides in
an inert medium**

   A detailed study of the thermal degradation of a commercial polyimide
product — "H-film" — based on gravimetric and mass-spectrometric
measurements was carried out by Bruck /45—49/. He showed that the
"H-film" is made of polypyromellitimide of the structure

   The samples studied were films 25μ thick, both as-cast and additionally
purified. The purification consisted in soaking the film at room temperature
for 72 hours in dimethylformamide, 24 hours in water and 24 hours in
ethanol in succession and then drying for 12 hours at 120° in vacuo. This
purification resulted in partial removal of impurities and macromolecular
chains with a large proportion of noncyclized amido-acid links.

Figure 20 gives the results of determinations of the loss in weight with time for purified and unpurified "H-film" at different temperatures in vacuo. The loss in weight has been expressed in relative units, as follows:

$$\delta P = \frac{\Delta P}{P_0} \cdot 100\%.$$

where P_0 is the initial weight of the sample and ΔP is the decrease in weight. The zero time corresponds to the moment at which the given temperature has been established. The time required for the temperature to be thus established was usually $25-30$ minutes; during this time some loss in weight of the sample also occurred, but it did not amount to more than 8%.

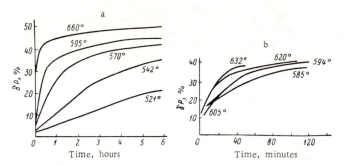

FIGURE 20. Thermogravimetric curves given by "H-film" at different temperatures (degradation in vacuo) /46 −48/.

a — unpurified "H-film"; b — purified "H-film."

It is seen from Figure 20 that the main loss in weight is incurred in the initial degradation period. Subsequently, the weight of the residual sample tends to some limiting value, which is different for the different experimental temperatures. At the highest temperature (660°) the unpurified film lost about 50% of its weight in 6 hours; the maximum weight loss for purified samples was about 40%.

The degradation proceeds at a variable rate. When the rate of weight loss $v = \frac{d(\delta P)}{dt}$ by the unpurified "H-film" (Figure 21a) is plotted as a function of the weight loss δP, the maximum corresponds to $\delta P = 10-20\%$. When the total weight loss attains $35-40\%$, the loss rate drops almost to zero. The descending segments of these curves are partly rectilinear. For the purified film (Figure 21b), the function consists of two rectilinear segments. According to Madorsky /79/, the linear segments represent the steady degradation process. It is considered that the deviations from linearity in the initial segments are caused by the presence of impurities, by the weak sites in the chains and other minor factors, the effect of which is significant at the initial stages of the degradation and then disappears. Extrapolation of the linear segments to $\delta P = 0$ gives the "apparent initial rate" of the degradation v_0. This rate for the purified film is about four times lower than for the unpurified film, under the same experimental conditions.

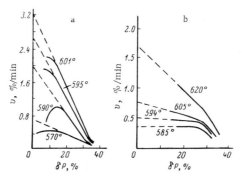

FIGURE 21. Rate of weight loss v as a function of the relative weight loss δP at different temperatures for the "H-film" (degradation in vacuo) /46—48/.

a — unpurified "H-film"; b — purified "H-film."

Using the Arrhenius' equation

$$k = A e^{-\frac{U}{RT}},$$

it is possible to find the activation energy U of the degradation process by plotting log v_0 against $\frac{1}{T}$, since the rate of weight loss is proportional to the rate constant of this reaction. This is shown in Figure 22. The logarithm of the apparent initial degradation rate, found by extrapolation

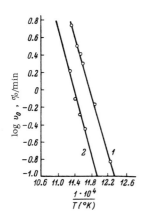

FIGURE 22. Logarithm of the rate of degradation of the "H-film" in vacuo as a function of reciprocal absolute temperature /46—48/.

1 — unpurified "H-film"; 2 — purified "H-film."

(cf. Figure 21), is plotted on the ordinate. The experimental points lie on a straight line both for the purified and the unpurified samples. The resulting values of the activation energy are similar in both cases — 74 and 73 kcal/mole for the purified and unpurified samples respectively. It may be pointed out that such values of the activation energy of the degradation reaction are also obtained for the unpurified sample if the rate of decomposition is measured by the maximum weight loss rate or some other parameter of weight loss (Figure 21a). The values for the purified sample are somewhat lower if the initial rate is determined by extrapolation of the second linear segments (Figure 21b).

It is seen from Table 7 that the activation energy of the degradation of polypyromellitimide is the highest of those given by polymers listed in /47/ for the data of /44, 79, 80, 103/.

The chemical mechanism of thermal degradation cannot be established on the strength of thermogravimetric data alone. More information can be gleaned from the analyses of the substances evolved in the course of the decomposition. It was found /46—48/ that the standard "H-film" contains large amounts of amido-acid links, which have not undergone ring closure during imidization. Their presence is indicated by 3400 cm^{-1} and 1680 cm^{-1} absorption bands, produced by valency vibrations of the N—H bond and carbonyl respectively. These bands are much weaker in the spectrum of the purified "H-film," but are clearly seen in the spectrum of the unpurified film and especially so in the spectrum of the sublimate which separates out of it in the amount of $2-3\%$ during the early stages of the degradation at $200-250°$.

TABLE 7. Activation energy of the process of thermal degradation in vacuo for a number of high polymers /47/

Polymer	u, kcal/mole
Polypyromellitimide ("H-film")	74
Polytrivinylbenzene	73
Polymethylene (linear)	72
Poly-α-methylstyrene	65
Polypropylene (linear)	58
Polystyrene	55
Polybenzyl	50
Polyisobutylene	49
Poly(methyl methacrylate)	30—52
Polycaprolactam	24—43

According to Bruck, the polymeric macromolecules of the industrial "H-film" contain structural units of at least three different types:

$$\left[-N\genfrac{}{}{0pt}{}{CO}{CO}R\genfrac{}{}{0pt}{}{CO}{CO}N-R'-\right]_n ; \quad \left[-N\genfrac{}{}{0pt}{}{CO}{CO}R\genfrac{}{}{0pt}{}{COOH}{CO-NH-R'-}\right]_m ;$$

$$(I) \qquad\qquad\qquad\qquad (II)$$

$$\left[-N\genfrac{}{}{0pt}{}{CO}{CO}R\genfrac{}{}{0pt}{}{CO-NH-R-}{CO-NH-R'-}\right]_p .$$

$$(III)$$

It is assumed that the content of type (II) structures in the purified film is much lower than in the unpurified, and that it is much lower in both kinds of films than the content of structures of type (I) $(m \ll n)$. Structures of type (III) are not very likely to be encountered. It is, however, difficult to assume that all these groupings may form part of the same polymeric chain in view of the experimental fact that the film can be purified.

In addition to amido-acid links, "H-film" may also contain a small amount of amines and traces of water. This is indicated, in particular, by the presence of absorption bands at $3450-3650 \text{ cm}^{-1}$.

Mass-spectrometric analysis of gaseous products of decomposition of the "H-film," condensed at the temperature of liquid nitrogen, indicates the evolution of large amounts of CO_2 and water (Table 8).

Purified film contains less CO_2 than does unpurified film. Table 8 does not contain data on the content of carbon monoxide which is condensed below the temperature of liquid nitrogen.

TABLE 8. Mass-spectrometric analysis of gaseous products of decomposition of "H-film" (600°C, 4 hours, $1 \cdot 10^{-3}$ mm Hg) /48/

Substance	Content, mole %	
	unpurified film	purified film
Carbon dioxide	38.0	25.4
Water	53.0	72.6
Ammonia	2.1	None
Hydrogen cyanide	5.6	"
Hydrocarbons	0.8	1.0
Aniline	0.05	0.05
Benzene	0.6	0.4
Phenol	0.04	0.2
Benzonitrile	0.06	0.1

The fact that the CO_2 content is much lower in the products of decomposition of the purified film indicates that the evolution of CO_2 is mainly due to the decomposition of impurities or amido-acid links. It was in fact shown by Straus and Wall /102, 103/ that pyrolysis of polyamides is accompanied by the evolution of a large amount of carbon dioxide.

According to Bruck, polyamido-acids are decomposed both as a result of the application of heat and as a result of hydrolysis, as follows:

where $R' = -\langle \rangle - O - \langle \rangle -$.

Hydrolysis of amido-acid groups may be produced by water sorbed by the polymer and held by hydrogen bonds. It is seen from the scheme that thermal and hydrolytic scission of the "impurity" amido-acid links should result in the evolution of larger amounts of carbon dioxide, as is actually the case.

The mechanism of decomposition of the main component of the "H-film"— pyromellitimide links — has also been deduced from the composition of the products evolved and elementary composition of the residue. Table 9 shows data of elementary analysis of the initial film and residue after decomposition at 600°.

TABLE 9. Elementary analysis of "H-film" and residue after heating (4 hours, 600°C, $1 \cdot 10^{-3}$ mm Hg) /48/

Element	Film		Residue	
	found, %	calculated, %	found, %	calculated, %
C	67.5	69.1	81.0	80.0
H	2.8	2.6	2.7	3.6
N	7.9	7.3	8.3	10.3
O	21.0	20.6	4.2	5.5

N o t e . Printer's errors in the original paper /48/ have been corrected.

In calculating the elementary composition it was assumed that the initial film was 100% imidized and that the carbonyl groups evolved during the degradation originated from the imide rings alone. Similar calculated data for the elementary composition are obtained if it is assumed that the initial film contains 80% imide and 20% amido-acid links by weight, that the degradation is accompanied by the evolution of four CO molecules per imide link and that the amido-acid is decomposed according to the scheme given above. These assumptions are in good agreement with the fact that the residue of the sample becomes richer in nitrogen and that its oxygen content decreases considerably.

The data in Table 8 were obtained as a result of heating in vacuo, accompanied by continuous evacuation of gases from the reaction chamber and trapping the products condensing above −195°. Experiments carried out in vacuo in an enclosed space /48, 69/ made it possible to analyze all or almost all products evolved during the degradation of the "H-film" (Table 10). It is difficult to make a quantitative comparison of the results obtained in /48/ and /69/, since the degree of decomposition was larger in the former case, while the samples may have consisted of different parts of the film. It is seen, nevertheless, that carbon monoxide was the main product in both cases.

On the strength of this finding and of the data of elementary analysis, and in view of the fact that the imide ring bonds are the least thermally stable in the polyimide, Bruck concluded that the decomposition of the polyimide part of the "H-film" consists in the disruption of these rings with separation of carbon monoxide and formation of the carbonized nitrogen-containing residue:

On the assumption that carbon dioxide is evolved only during the decomposition of polyamido-acid links and in view of the results given

in Table 10, it may be concluded that commercial "H-film" contains about
78% of polyimides and 22% of polyamido-acid. This is not in contradiction
with the data of elementary analysis (Table 9). Since the molecular weight
of the repeating unit of the polymer is 382, the loss of four CO groups per
repeating unit is equivalent to a weight loss of about 30%, calculated on
pure polyimide. If we allow for the presence of impurities, the experimental
loss in weight of about 40% (Figure 20) is in good agreement with the
postulated mechanism of thermal degradation in vacuo.

TABLE 10. Mass-spectrometric analysis of products of decompostion of unpurified "H-film" in vacuo
in an enclosed space at 610°C /48/ and 540°C /69/

Substance	Content		
	mole % /48/	wt. % /48/	mole % /69/
Hydrogen	9.0	0.64	26
Methane	1.1	0.63	–
Ammonia.	1.2	0.73	traces
Water	7.6	4.90	1.2
Hydrogen cyanide	1.3	1.26	1.2
Carbon monoxide	60.5	60.73	58.1
Carbon dioxide	19.0	29.97	35.1
Benzene	0.3	0.84	0.7
Phenol	0.02	0.07	–
Benzonitrile.	0.06	0.22	0.5

The residue obtained as a result of advanced degradation (pyrolysis) of
the "H-film" in vacuo at elevated temperatures constitutes about 60% of
the weight of the initial sample. It was shown by Bruck /45, 49/ that this
residue had a high density and a high electrical conductivity. Figure 23
shows the values of specific resistance and density of samples of "H-film"
50μ thick held for 1 hour in vacuo at different temperatures, beginning
from 620°. The resistivity decreases sharply and the density of the
pyrolysis product increases with the increase in the experimental
temperature. At earlier stages in the pyrolysis (below 620°), on the
contrary, the density was observed to decrease from 1.426 to 1.330 g/cm³.
At this stage the bulk of the volatile products is evolved and the
residue contains high concentrations of free radicals — average concentration attains 10^{19} spins/g—as indicated by EPR method. The residual
concentration of free radicals depends on the duration of the pyrolysis at
the given temperature. The dependence of the concentration on the time
usually displays a maximum (Figure 24). The height of the maximum
is determined by the temperature of the pyrolysis; it is highest at 620°.
When the experimental temperature increases, there is a sharp drop
in the residual concentration of free radicals.

According to Bruck, the thermal transformations of "H-film" take place
in two stages. The first stage — thermal degradation — consists in the
scission of C—N and C—C bonds in imide rings, which is accompanied by
the evolution of carbon monoxide. The nuclei with the free bonds thus
formed partly recombine with each other (see also pp. 35, 36). The EPR
signal originates from the unreacted part of free radicals. However, such

43

systems still cannot ensure the overlap of electronic π-orbits which would be sufficient for a continuous conductance zone to appear. The second stage — pyrolysis — is distinguished by an almost complete absence of any loss in weight, an increase in density and a sharp decrease in the resistivity of the sample. This stage involves a complete rearrangement of the internal structure of the material, consisting in spatial polycondensation of aromatic rings. This is accompanied by a marked increase in unsaturation and the appearance of electric conductance. The decrease in paramagnetic absorption which is observed at this stage is due to the recombination of free radicals.

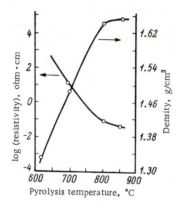

FIGURE 23. Variation of density and resistivity of "H-film" after pyrolysis in vacuo at different temperatures during 1 hour /45/.

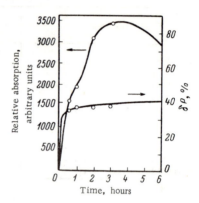

FIGURE 24. Variation of paramagnetic absorption and weight at 20°C of a sample of "H-film" during pyrolysis in vacuo at 620°C /45/.

It is probable that the ideas of Bruck are a correct representation of the main trends of thermal transformations undergone by polypyromellitimide in vacuo. Nevertheless, several points are as yet unproved and contradictory.

It is noteworthy, in the first place, that the degradation is always accompanied by the evolution of large amounts of water (Tables 8 and 10). Now the mechanism proposed for the decomposition of the impurity molecules of polyamido-acids (p. 41) involves absorption rather than evolution of water. The elimination of oxygen solely as CO which originates from direct decomposition of imide links also precludes the formation of water. The evolution of some water could be easily explained by the cyclization of amido-acid links present as impurities during the destructive heating, but this has not been postulated by Bruck.

Further, if we assume that the evolved CO_2 originates exclusively from the decomposition of polyamido-acid links, Bruck's scheme implies the evolution of 1 mole of benzene for each 4 moles of CO_2. It will be seen from Tables 8 and 10 that the amount of benzene in fact evolved is several times lower than that. Thus, the reaction mechanism proposed to describe the thermal degradation is not in agreement with the actual contents of water and benzene in the decomposition products.

Moreover, it has not been conclusively established that oxygen is evolved as a result of decomposition of imide rings as CO only, without forming any other compounds. This could be readily checked by working with samples which are known not to contain amido-acid or amide groups. It has been shown /100/ that such polyimides can in fact be prepared.

As regards the final stage of thermal transformation — pyrolysis — the following may be said. It is beyond doubt that heating at high temperatures results in a complete rearrangement of the internal structure of polypyromellitimide. This is demonstrated by the observed major changes in the density and conductivity. It is however difficult to say whether or not the actual mechanism of these transformations is in fact represented by the scheme proposed by Bruck. The nature of the free radicals has not been established with certainty, and it may be assumed that they are formed as a result of decomposition of the imide rings, i.e., at the thermal degradation stage. This is in agreement with /6/ and with the results of Bruck himself. If the concentration is 10^{19} spins/g, one radical corresponds to about 1000 broken bonds. It follows that the formation of free radicals is accompanied by their intense recombination even prior to the pyrolysis stage. According to Bruck, this stage consists in the formation of polycyclic groupings, the presence of which would thus be conspicuous during the first stages of the process. This would be accompanied by the appearance of new IR absorption bands, which has in fact not been observed even in the advanced stages of the transformation.

The special value of Bruck's work consists in his rich factual material which can be seen to lead to the main features of the mechanism of thermal degradation of polyimides in vacuo and to point the way to further studies.

Thermooxidative degradation of polypyromellitimides

Bruck /46—48/ also carried out experimental studies on the thermal degradation of the "H-film" in the air. As distinct from the degradation in vacuo, the weight loss in oxygen-containing atmosphere may be as high as 95% (Figure 25). The degradation process proceeds at a more intense rate at much lower temperatures. The rate of the process varies by a factor of not more than 2 — 3 up to complete decomposition, whereas in vacuo the decrease factor is 10 or more for a smaller range of loss in weight (cf. Figures 20, 21). The activation energies of thermooxidative degradation, calculated from the maximum rates of loss in weight (which occur at $\delta P \sim 50\%$) are 33 and 31 kcal/mole for purified and unpurified samples of "H-film" respectively. This, in conjunction with the fact that the sample is decomposed to gaseous products, shows that the thermal degradations in vacuo and in air proceed according to different mechanisms. Unfortunately, no analytical data on the products of degradation of the "H-film" in the air are given in /46—48/

Scala and Hickam /96/ made a very detailed and careful study of thermooxidative degradation working on finely dispersed powders of specially prepared polypyromellitimide

(I)

and polypyromellitamidoimide

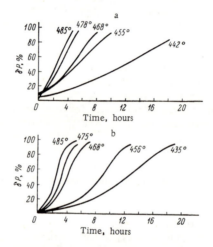

The authors /96/ followed the kinetics of the variation of sample weight and at the same time also the variation in the gas composition (O_2, CO_2 and CO) of the atmosphere in which the samples were immersed, by means of manometric and mass-spectrometric measurements. The reaction between the polymers and oxygen was also studied by following the variation of the composition of the oxygenated atmosphere enriched in O^{18}. An attempt was then made to establish the molecular mechanism of the thermooxidation of polyimides from the results of quantitative juxtaposition of the variation in sample weight and the variation in the concentration and composition of the gases.

FIGURE 25. Thermogravimetric curves of "H-film" (decomposition in the air at different temperatures) /46—48/.

a—purified "H-film"; b—unpurified "H-film."

Samples were prepared by film casting, drying, grinding and heating the resulting powders in vacuo at 150°. It was found that in order to obtain reproducible results, the samples had to be conditioned (outgassed) for several hours in vacuo or in helium at 300—400°. The conditioning resulted in the liberation of residual solvent, adsorbed water and low-molecular polymer fractions by the samples. The stability of the structure and the composition of the samples were controlled during the experiment by way of IR spectra and elementary analysis.

The main features of the interaction of polymers with oxygen may be illustrated by the results of experiments conducted on samples of poly-imide I at 300° in an atmosphere enriched in O^{18}. Sample I, which had been outgassed in vacuo between 100 and 300° by being held for one hour each at temperatures rising by 30° each time, was placed in a platinum chamber containing a mixture of 50% argon and 50% oxygen (90% O_2^{18} + 10% O_2^{16}) at 300°. Analyses of gas samples were used to plot kinetic curves of oxygen absorption and evolution of carbon dioxide of different isotopic compositions (Figure 26). It is seen that after two hours' heating the rates of variation of gas concentrations become constant. The rate of evolution of CO_2 was 0.018 moles/mole repeating unit in one hour (mole/mole hour), while the rate of O_2 absorption was 0.022 moles/mole·hour. It is thus easily calculated that the evolution of CO_2 results in a decrease in the sample weight at the rate of 0.00206 grams/gram·hour, while the absorption of O_2 results in a weight increase at the rate of 0.00185 grams/gram·hour. It follows that the resulting rate of decrease in the weight of the sample (by difference) is 0.00021 grams/gram·hour. In the case of the sample outgassed in vacuo for $2\frac{1}{2}$ hours at 300°, the stationary rate of weight decrease in the air at 300° was 0.000286 grams/gram·hour.

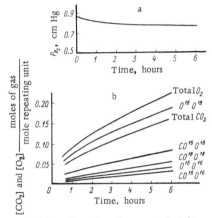

FIGURE 26. Absorption of oxygen and evolution of carbon dioxide by the outgassed sample of polyimide

at 300°C in a mixture of argon and oxygen enriched in O^{18} /96/.

a — variation of the partial pressure of oxygen in the course of the experiment; b — variation in the content of the gases analyzed.

A comparison of Figures 26 and 27 shows that at least 75% of the total weight loss is connected with changes in the composition of the gaseous atmosphere — absorption of oxygen and evolution of carbon dioxide. The high rates of loss of carbon as CO_2 (0.00056 g/g·hr), evolution of CO_2 (0.00206 g/g·hr) and absorption of O_2 (0.00185 g/g·hr) by the polymer show that the low overall rate of weight loss ($0.000286\ g/g \cdot hr$) is not the result of the chemical inertness of the polymeric substance, but, on the contrary, of its intense reaction with oxygen, The fact that the rate of weight loss of carbon as CO_2 alone is twice as high as the overall rate of weight loss (0.00056 as against $0.000286\ g/g \cdot hr$) shows that the interaction of the polymer with oxygen cannot be limited to the cleavage of large individual parts of the macromolecules. Such cleavage often takes place in polymers with links which are weak with respect to oxygen. Even though 90% of the oxygen absorbed by the polymer consists of O^{18}, the bulk of the evolved CO_2 contains the O^{16} isotope. In order to identify more readily the trans- formations undergone by oxygen, the data of Figure 26 have been recalculated to give the absorbed and evolved O_2^{16} and O_2^{18} (Figure 27). It is seen from Figure 27 that when oxygen is absorbed by the polymer, the O_2^{18}/O_2^{16} ratio in the gaseous mixture remains constant throughout and is equal to the initial concentration of the gas mixture ($90\%\ O_2^{18} + 10\%\ O_2^{16}$). This means that there is no isotopic exchange between the gas medium and the polymer. It may be calculated from the amount of O_2^{18} oxygen which is absorbed and evolved as CO_2 that the bulk of it is retained by the polymer. If it is assumed that the position is similar as regards the O_2^{16} oxygen of the gas mixture, it will be found that the bulk of the O^{16} isotope evolved as carbon dioxide originates from the polymer. The total amount of the evolved oxygen of all kinds is less than its total absorbed amount. In advanced oxidation (see below) practically all the absorbed oxygen is returned to the atmosphere as CO and CO_2.

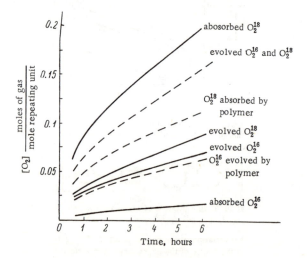

FIGURE 27. Absorption and evolution of oxygen isotopes (as CO_2) under conditions of Figure 26 /96/.

Continuous curves represent experimental data; broken lines represent calculated data.

Thus, it was conclusively established in these experiments that thermo- oxidative degradation of polyimides is accompanied by an intense interaction between the polymer substance and oxygen, the final result of which is mainly formation of simple gaseous products. However, since the evolved CO_2 only was recorded, it was not possible to establish a full material balance between oxygen and the reaction products or the reaction mechanism.

More definite results in this respect were obtained by the authors /96/ for prolonged oxidation of polyamidoimide II, during which the changes in weight, and concentrations of oxygen, carbon dioxide and carbon monoxide were recorded at the same time. The most nonambiguous results were obtained on sample II which had been outgassed at 400° and tested at 360° in an atmosphere of 60% argon and 40% oxygen. The absorption of oxygen, evolution of carbon dioxide and carbon monoxide (Figure 28) and the change in sample weight (Figure 29) were quantitatively determined. The rate of these processes remained practically constant throughout the experiment, which lasted for more than 100 hours. By the end of the experiment the total loss in weight of the sample was about 28% of the initial weight. Using the data of Figure 28 it was possible to establish the fraction of the overall weight loss due to carbon evolved as CO and CO_2. It can be seen (Figure 29a) that out of the 28% loss in weight at the end of the experiment, about 20% is due to carbon, i.e., some 70% of the total loss in weight. A molecule of polymer II contains 14.6% carbon in the form of carbonyl groups (11.7% in imide + 2.9% in amide), while the total carbon content is 67.5%. The curve giving the carbon weight loss of the sample throughout the experiment is similar to the curve corresponding to 67.5% of total weight loss which may be calculated from the experimental weight measurements (Figure 29a). The authors /96/ concluded that the CO and CO_2 evolved during the thermooxidative degradation of polyimide do not originate from carbonyl groups only, but also from all the other atoms in the macromolecule. It also follows that advanced thermal oxidation involves a full destruction of each monomeric unit. That the $C = O$ groups are not the main source of the gaseous carbon-containing products evolved (at least of CO_2) is additionally confirmed by the calculated graph of loss in weight related to the weight of the sample at any given moment, produced by the evolution of the gaseous products (Figure 29b). In fact, these curves pass, without breaks, through points corresponding to the loss of four and five $C = O$ groups by all monomer units. At the same time the $C = O$ groups, and especially the amide groups, are the primary points of attack on the macromolecular chain by oxygen, which is followed by complete destruction of the cleaved fragment.

To summarize, the main results obtained during the 105-hour experiment at 360° may be stated as follows. The carbon content of the products evolved is the same as that in the polymer. The elementary compositions of the residue and of the starting polymer are the same and are close to the theoretical, within the limits of experimental accuracy. The IR spectra of the residue throughout the experiment are also similar to that of the initial sample. The rates of evolution of volatiles and the loss in weight are constant throughout the duration of the experiment and thus the absolute number of sites at which oxygen interacts with the polymer is constant as well.

These facts indicate that the oxidation reaction proceeds from the ends of the macromolecules (the number of which is constant) and leads first to the cleavage of large fragments of the monomeric unit (e.g., at the amide grouping) and then to their complete destruction with the formation of simple gaseous products.

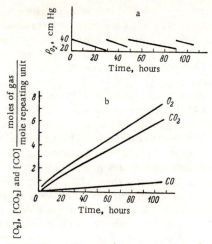

FIGURE 28. Oxidation of outgassed sample of polypyromellitamidoimide

in a mixture of 60% argon and 40% oxygen at 360°C /96/.

a — change of partial oxygen pressure with time; b — change of oxygen absorption and CO_2 and CO evolution with time.

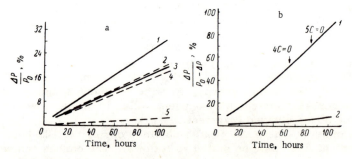

FIGURE 29. Loss in weight during the oxidation of polypyromellitamidoimide /96/

a — temporal variation of weight loss, calculated on the initial sample weight: 1 — total loss; 2 — loss in carbon as CO and CO_2; 3 — 67.5% of total loss; 4 — carbon lost as CO_2; 5 — carbon lost as CO; b — temporal variation of weight loss, calculated on the weight of sample at the given moment: 1 — carbon lost as CO_2; 2 — carbon lost as CO.

These conclusions are confirmed, in general, by experiments at 400° which resulted in an even more advanced oxidation of polyamidoimide II (loss in weight 80% of the initial). It was also found that only carbon and

50

hydrogen are oxidized to the gaseous products. About 50% of the nitrogen is retained in the solid residue, carbon monoxide is formed mainly as a result of cleavage of C=O groups, while CO_2 is evolved as a result of oxidation of phenyl nuclei.

It would appear that the molecular mechanism of thermal oxidation of polyimides proposed by Scala and Hickam /96/ is in fact the main mechanism, but not the only one. This is evident even from the fact that a complete material balance based on it can be drawn up only for the medium values of loss in weight on oxidation.

Comparison of the results obtained by studying the degradation of poly-pyromellitimides in inert and oxygen-containing media shows that in the latter case the chemical changes affect all groups and elements of the polymer chain, apparently excepting nitrogen. During purely thermal degradation, on the other hand, the chemical changes affect almost exclusively the most highly reactive groupings.

Effect of chemical structure of polyimides on their thermal and thermooxidative stability

The stability of polyimides to high temperatures is obviously determined, first and foremost, by their chemical structure. Data on the dependence of thermal stability on the chemical structure are very important both as indications of the mode of preparation of other thermostable polyimides, and in the choice of the most promising individual products. Most information available in the literature refers to polyimides prepared from pyromellitic dianhydride and different diamines, i.e., to polypyromellitimides.

TABLE 11. Thermal stability of polypyromellitimides by the DTA method /86/

Polymer	R'	In nitrogen		In the air
		incipient decomposition temperature, °C	peak temperature, °C	incipient oxidation temperature, °C
I	–⟨◯⟩–	500	610	450
II	(naphthalene ring)	460 (540)	590	300
III	–⟨◯⟩–⟨◯⟩–	510	615	410
IV	H₃C, CH₃ –⟨◯⟩–⟨◯⟩–	490	540	330
V	–⟨◯⟩–O–⟨◯⟩–	490	595	400
VI	⟨◯⟩–SO_2–⟨◯⟩–	420	485	300
VII	–⟨◯⟩–CH_2–⟨◯⟩–	480	550	230
VIII	–⟨◯⟩–$(CH_2)_2$–⟨◯⟩–	470	580	200
IX	–⟨◯⟩–$C(CH_3)_2$–⟨◯⟩–	400 (450)	430 (485)	320
X	–$(CH_2)_4$–	370	430	290

Nishizaki and Fukami /86/ studied the thermal degradation of a number of polypyromellitimides containing different groupings in the capacity of the R' radical in the diamine component (Table 11, Nos. I—X). The experiments have been performed in the air and in nitrogen at temperatures up to 700°. The method used was differential thermal analysis (DTA). Polypyromellitimide samples were in the form of finely ground powders. Aluminum oxide was used as etalon. The temperature was raised at the rate of 10°/min.

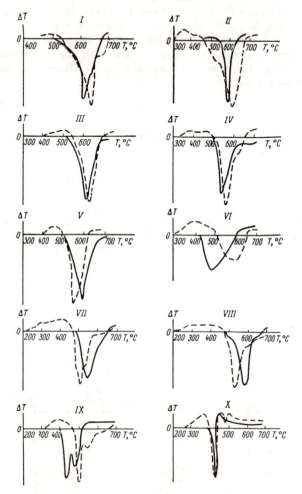

FIGURE 30. DTA diagrams of polypyromellitimides /86/.

Roman numerals correspond to polymer numbers in Table 11.
Continuous curves — heating in nitrogen, broken curves —
heating in the air.

The DTA diagrams taken in /86/ are represented in Figure 30. The cell temperature has been plotted on the abscissa, while the difference between the temperatures of the sample and of the etalon has been plotted on the ordinate (the scale has not been indicated). A positive temperature difference corresponds to the evolution of heat by the sample, while a negative difference corresponds to the absorption of heat by the sample.

The DTA diagrams indicate in all cases that the main processes are accompanied by the absorption of heat. This may be due to the decomposition or sometimes to the melting of the polymer. When the experiments are conducted in the air, a small heat evolution usually occurs at first, and is then followed by heat absorption. It would appear that the evolution of heat is connected with oxidation.

In most cases there is only one sharp peak of the absorbed heat, and in only one case (polymer VI) does the peak spread over a wide temperature range. The DTA diagram of polymer IX clearly shows two absorption peaks. It is possible that the thermal degradation of the polymer is preceded in this case by fusion. The other polyimides studied show no fusion peaks. Table 11 shows the temperatures which correspond to the incipient intense heat absorption and the peak temperatures of the experiments carried out in nitrogen. If we assume that the incipient decomposition temperature (incipient heat absorption temperature) is an index of thermal stability, the radicals R' (and of the corresponding polypyromellitimides) can be arranged in the following decreasing sequence of thermal stability:

$$-\text{C}_6\text{H}_4-\text{C}_6\text{H}_4-\ >\ -\text{C}_6\text{H}_4-, \quad \text{(naphthalene)}\ >\ -\text{C}_6\text{H}_4-\text{O}-\text{C}_6\text{H}_4-,$$

$$-\text{C}_6\text{H}_2(\text{H}_3\text{C})-\text{C}_6\text{H}_2(\text{CH}_3)-\ >\ -\text{C}_6\text{H}_4-\text{CH}_2-\text{C}_6\text{H}_4-\ >$$

$$>\ -\text{C}_6\text{H}_4-(\text{CH}_2)_2-\text{C}_6\text{H}_4-\ >$$

$$>\ -\text{C}_6\text{H}_4-\text{SO}_2-\text{C}_6\text{H}_4-\ >\ -\text{C}_6\text{H}_4-\text{C}(\text{CH}_3)_2-\text{C}_6\text{H}_4-\ >\ -(\text{CH}_2)_6-.$$

The radicals $-\text{C}_6\text{H}_4-\text{SO}_2-\text{C}_6\text{H}_4-$ and $-\text{C}_6\text{H}_4-\text{C}(\text{CH}_3)_2-\text{C}_6\text{H}_4-$ may change places if in the latter case the decomposition temperature is assumed to be indicated by the second and not by the first peak, as was in fact done in /86/. Certain conclusions may be arrived at on the strength of this sequence. The highest thermal stability is displayed by polypyromelliti-mides obtained from purely aromatic diamines — polymers III, I (R' = $-\text{C}_6\text{H}_4-\text{C}_6\text{H}_4-$ and $-\text{C}_6\text{H}_4-$). They begin to decompose at 500° and above. This group also seems to include the polyimide II $\left(\text{R'} = \text{(naphthalene)}\right)$, which begins to absorb heat at 540° (the temperature of 460°, which is given by the authors /86/ and which is found in Table 11 corresponds to incipient evolution of heat, cf. Figure 30). This is also in agreement with subsequent data.

The introduction of substituents into the benzene ring (polymer IV) and the introduction of various groupings between the benzene nuclei of the diamine component (especially sulfone and isopropylidene) results in a decrease in the decomposition temperature. Polyimides based on aliphatic diamines are the least resistant to heat.

Thermal stability in the presence of oxygen may be estimated from the temperature corresponding to incipient evolution of heat (last column of Table 11). If this is done, it is found that the most oxidation-resistant are polypyromellitimides obtained from aromatic diamines in which the phenyl nuclei are interlinked in the para-position — polymers I and III. The ether bond (polymer V) somewhat reduces the stability to thermal oxidation. The presence of methyl substituents in benzene rings, different groups between phenyl rings in R' and benzene rings in meta-position markedly reduces the oxidation temperature. The sequence of radicals R' of the diamines in the order of decreasing resistance to thermal oxidation is as follows:

If this sequence is compared with the decreasing sequence of thermal stability of radicals R' given above, it is seen that the thermal and thermo-oxidative stabilities of numerous groupings are quite different.

Some of the polymers studied in /86/ by the DTA method (V, VII and X in Table 11) were further investigated by thermogravimetric and IR spectroscopic methods /87/. These methods and similar experimental conditions were used by Nishizaki and Fukami /88/ in their study of two other polypyromellitimides, the R' radicals of the diamino component of which had the following structure:

Fine polyimide powders were again used as samples. Thermogravimetric determinations were performed in the air and in nitrogen; the heating rate was 5 degrees/minute. The IR spectra of the initial samples, decomposed samples and products of decomposition were taken at room temperature.

Figure 31a represents the thermogravimetric curves of these polypyro-mellitimides obtained in nitrogen. It is seen that the shape of the curves is approximately the same in all cases. Figure 31b shows the rate of decrease in weight as a function of temperature, calculated from the curves. In all cases there are temperature ranges in which the rate of decrease in weight was at maximum. A change in the structure of the diamine had a marked effect on the temperature of incipient major weight loss. If these temperatures are compared with the incipient decompostiion temperatures obtained by the DTA method (Table 11, polymers V, VII, X), the latter are seen to be always somewhat higher. This is mostly due to the fact that in the DTA method the heatup rate was higher, since in both cases the relative thermal stability sequence of the polymers remains the same.

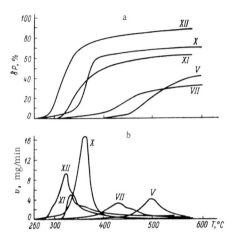

FIGURE 31. Thermogravimetric curves (a) and weight loss rates as functions of temperature (b) for polyimides V, VII, X, XI, XII (Table 11).

Heatup rate 5 degrees per minute. Experiments performed in nitrogen /87, 88/.

The decompostion of polymers X, XI, and XII

$$\left(R' = -(CH_2)_6-, \ -CH_2-\bigcirc-O-\bigcirc-CH_2-, \right.$$
$$\left. -CH_2-Si(CH_3)_2-O-Si(CH_3)_2-CH_2- \right)$$

begins at the lowest temperatures (280—320°). These polymers also display high decomposition rates. For example, the decomposition rate of polymer X, obtained from an aliphatic diamine, is 16.7 mg/min for an initial sample weight of 100 mg. Polymers X—XII also display large overall weight losses at high temperatures. This group of polymers is distinguished by the presence of a direct bond between the imide ring and the methylene link $\left(-CH_2-N\big< \right)$.

Polymers VII and V

$$\left(R' = -\bigcirc-CH_2-\bigcirc-and-\bigcirc-O-\bigcirc- \right)$$

begin to decompose at much higher temperatures (380—450°). The maximum decomposition rates also correspond to higher temperatures

55

(436 and 500°) and the size of the maxima is small. The lowest rate of decomposition is shown by polyimide VII — 3.5 mg/min. Its decomposition takes place in a wide temperature range.

If the decomposition takes place in the air, the weight loss occurs at lower temperatures than in nitrogen (Figure 32a). Some of the polymers — e.g., VII and X — display two stages of decomposition, judging by the shape of the curves representing the rate of weight loss as a function of temperature (Figure 32b). The first marked rise in the decomposition rate is apparently due to the effect of oxygen. The main rate peak corresponds to the same temperature as the decomposition peak in nitrogen. It would appear that at this stage, both in the presence and in the absence of oxygen, the decomposition proceeds according to the same mechanism, while the effect of oxidation is most marked at earlier stages.

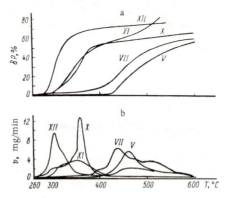

FIGURE 32. Thermogravimetric curves (a) and weight loss rates as functions of temperature (b) for polyimides V, VII, X, XI, XII (Table 11).

Heatup rate 5 degrees per minute. Experiments performed in air /87, 88/.

In polymers V and XII this effect is practically absent. However, their weight loss rate vs. temperature curves shift by 40 –50° towards lower temperatures in experiments conducted in the air.

In a study of polyimides decomposition by IR spectroscopy /87, 88/, the polymers were degraded at constant temperatures for given periods of time. Degradation of polyimides V, VII and X was effected at temperatures corresponding to incipient intense decomposition (incipient rise in weight loss rate vs. temperature curve). The experiments were performed both in the air and in an inert gas. Since the residue formed as a result of the decomposition of polymer X was partly soluble in dimethylacetamide, the soluble and the insoluble fractions were studied separately.

The degradation temperature of polymers XI and XII was 300°. The experiments were performed in nitrogen atmosphere only. Spectra were taken of the solid residue and of the volatile decomposition products, condensed at the temperature of dry ice.

The group of polymers X, XI and XII is distinguished by the fact that the spectra of their decomposition products contain an absorption band at 3270 cm^{-1}, which is typical of the N—H bond of ring imides, e.g., phthalimide. In the case of polymer X this band is particularly distinct in the spectrum of the soluble fraction of the residue from decomposition both in nitrogen and in the air; in the case of polymers XI and XII the band appears in the spectrum of volatile decomposition products.

In addition, the spectrum of the soluble fraction of the residue of polyimide X showed a strong decrease in the intensity of the band at 2924 cm^{-1} which is produced by vibrations of methylene groups. Spectrum of volatile products of sample XII displays an increased intensity of the band at 1266 cm^{-1} which corresponds to symmetrical deformation vibrations of methyl groups. This band is typical of methylsiloxane /114/. As an example, the IR absorption spectra of polymer X and its decomposition products during heating in nitrogen are shown in Figure 33.

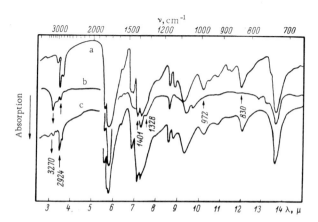

FIGURE 33. IR absorption spectra of polypyromellitimide X (R' = —(CH$_2$)$_6$—) and its residue after decomposition for 5 hours at 320°C in nitrogen /87/.

a — initial sample; b — soluble fraction of residue; c — insoluble fraction of residue.

The inspection of the spectra of polymers in this group shows that the weakest links in polyimides X, XI and XII are the radicals R', which are disrupted, in the first place, at the N—CH$_2$ bonds. The pyromellitimide nucleus proves to be more stable. During the disruption of the polymer chain it can form volatile and readily soluble compounds containing pyromellitimide and low-molecular chain fragments, with terminal links of the phthalimide type

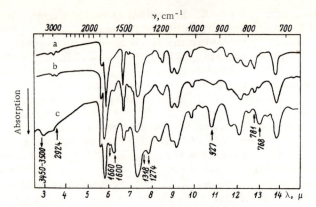

FIGURE 34. IR absorption spectra of polypyromellitimide VII

$\left(R' = -\langle\!\!\!\!\bigcirc\!\!\!\!\rangle - CH_2 - \langle\!\!\!\!\bigcirc\!\!\!\!\rangle - \right)$ and its residue after decomposition /87/.

a — initial sample; b — residue (5 hours, 420°C, in nitrogen); c — residue (5 hours, 360°C, in the air).

The distillation products also include compounds formed as a result of decomposition of the diamine component, in particular methylsiloxane formed by decomposition of sample XII. Polymeric chains not degraded to a significant extent remain behind in the insoluble and nonvolatile solid residue. Accordingly, the spectra of the solid residue do not differ much from those of the initial polymer.

The solid residue from the decomposition of polymers V and VII

$\left(R' = -\langle\!\!\!\!\bigcirc\!\!\!\!\rangle - O - \langle\!\!\!\!\bigcirc\!\!\!\!\rangle - \text{ and } -\langle\!\!\!\!\bigcirc\!\!\!\!\rangle - CH_2 - \langle\!\!\!\!\bigcirc\!\!\!\!\rangle - \right)$ in an atmosphere of

nitrogen did not contain soluble products and its spectra did not differ much (Figures 34a and 34b) from those of the initial polymer.

More significant changes in the spectra occur when the sample is degraded in the air. The spectrum of the solid residue of polyimide VII (Figures 34a and 34c) had more intense 3500, 1660, and 927 cm^{-1} and less intense 2924 and 781 cm^{-1} bands.

The thermooxidative degradation of this polyimide would appear to proceed

in a manner similar to that of polybenzyl $\left[-CH_2 - \langle\!\!\!\!\bigcirc\!\!\!\!\rangle - \right]_n$. It has been shown /74/

that the thermal oxidation of polybenzyl proceeds according to the scheme:

$$-\langle\!\!\!\!\bigcirc\!\!\!\!\rangle - CH_2 - \langle\!\!\!\!\bigcirc\!\!\!\!\rangle -$$

$$\downarrow O_2 \quad \Delta$$

$$-\langle\!\!\!\!\bigcirc\!\!\!\!\rangle - \underset{\underset{O-OH}{|}}{CH} - \langle\!\!\!\!\bigcirc\!\!\!\!\rangle -$$

$$-\langle\!\!\!\!\bigcirc\!\!\!\!\rangle - \underset{\underset{O}{\|}}{C} - \langle\!\!\!\!\bigcirc\!\!\!\!\rangle - + H_2O$$

$$-\langle\!\!\!\!\bigcirc\!\!\!\!\rangle - \underset{\underset{H}{|}}{\overset{\overset{OH}{|}}{C}} - \langle\!\!\!\!\bigcirc\!\!\!\!\rangle - \rightarrow \langle\!\!\!\!\bigcirc\!\!\!\!\rangle - \underset{\underset{O}{\|}}{\overset{\overset{OH}{|}}{C}} + \langle\!\!\!\!\bigcirc\!\!\!\!\rangle$$

2211

It may be assumed that polymer VII behaves in the same manner. In fact, the decrease in the intensity of the 2924 cm^{-1} band may be assumed to be the result of conversion of $-\overset{\text{H}}{\underset{\text{H}}{\text{C}}}-$ groups to $-\overset{\text{OH}}{\underset{\text{H}}{\text{C}}}-$ groups. This may be accompanied by hydrogen bond formation by the hydroxyls, with consequent appearance of the 3500 cm^{-1} band. The presence in the residue of benzo-

phenone type structures $\langle \bigcirc \rangle - \overset{}{\underset{\text{O}}{\text{C}}} - \langle \bigcirc \rangle$ is demonstrated by the presence

of the 1660 cm^{-1} band, which corresponds to valency vibrations of the C = O group of the diaryl ketone. The 927 cm^{-1} absorption band is also comprised in the spectrum of pure benzophenone.

According to the scheme given above the degradation of polyimide VII should be accompanied by the formation of carboxyl acid and thus also by the appearance of an absorption bands at 1700 cm^{-1} due to the C = O in the carboxyl group. The appearance of these bands was not noted in the spectra. The authors /87/ assume that these bands are masked by the strong absorption of imide groups in the same wavelength.

The decomposition of polymer V

$$\left(R' = -\langle \bigcirc \rangle - O - \langle \bigcirc \rangle - \right)$$

in the air was accompanied by the appearance of an absorption band at 3450 — 3500 cm^{-1} which was attributed /88/ to hydroxyl groups, and by some weakening in the 1244 cm^{-1} band (aromatic ether bond) relative to the absorption by the 1380 cm^{-1} band (cyclic imide C—N bond). This indicates that there is a decrease in the concentration of diphenyl ether

bonds $\left(\langle \bigcirc \rangle - O - \langle \bigcirc \rangle \right)$ as a result of thermal degradation in the air.

Unlike X, XI, and XII, the spectrum of polymer V has no absorption band at 3270 cm^{-1} which originates from the formation of imide N—H bonds, i.e., phthalimide and diimide type structures are not formed.

Data given in /86 —88/ justify certain general conclusions on the connection between the thermal stability of polypyromellitimides and their chemical structure. The pyromellitimide nucleus is a thermostable formation, apparently of a stability comparable to that of the most resistant diamines. If we compare the thermostabilities of different polypyromel-litimides (Table 11), we note that the decomposition temperature increases up to a certain limiting value, which is of the order of 500°. This is so even if the decomposition temperature of the diamine radicals is much higher. The stabilities of the diamines may be estimated from Table 12, which shows the decomposition temperatures of a number of model compounds /64/. The structure of many such compounds is that of diamine radicals mentioned above which are used in the synthesis of polypyromel-litimides. Thus, the thermal stability of polypyromellitimides is affected only by those diamine radicals which decompose below 500°. If the diamine radical is decomposed at a higher temperature, say, in the case

of R' = $-\langle \bigcirc \rangle - \langle \bigcirc \rangle -$, $-\langle \bigcirc \rangle - O - \langle \bigcirc \rangle -$, its presence as the constituent

of the polyimide will not raise the decomposition temperature of the latter. Thus, in the most highly heat-resistant polyimides, the element which determines their thermostability is not the diamine radical, but the pyromellitimide nucleus. Since the most stable pyromellitimides have a decomposition temperature of ~500°, this temperature may be considered as the stability limit of the pyromellitimide nucleus.

TABLE 12. Decompostion temperatures of certain low-molecular aromatic compounds /64/

Compound	Structure	Decomposition temperature, °C
Benzene.		593
Diphenyl		543
Diphenyl ether.		538
Triphenylamine		500
Quaterphenyl		494
Diphenyldiphenoxy-silane		482
Benzophenone		481
N,N'-Tetraphenyl-p-phenylene diamine . . .		458
Diphenylmethane.		455
Polyphenyl ether		440

Note. Decompostion temperatures were defined as temperatures at which the vapor pressures over the samples increased more rapidly than by 1%/hour.

In the studies of Bruck /45—49/ the polyimide with R' =

was decomposed at 500° and above. This is probably why the decomposition products included substances which could have been formed only as a result of disruption of the imide ring. On the other hand, low-molecular compounds of the diimide type were noted /87, 88/ in the volatile products formed by decomposition of polymers with a low thermal stability, which had

$$R' = -(CH_2)_6-, \quad -CH_2- \quad -O- \quad -CH_2-,$$

$$-CH_2-\underset{CH_3}{\overset{CH_3}{Si}}-O-\underset{CH_3}{\overset{CH_3}{Si}}-CH_2-.$$

Accordingly, in this case the imide rings had remained intact.

60

The least thermostable polypyromellitimides are those obtained from diamines containing methylene links, the position of such links being very important. This can be seen by comparing the thermostabilities of polymers in which $R' = -CH_2-\langle\bigcirc\rangle-O-\langle\bigcirc\rangle-CH_2-$ and $-\langle\bigcirc\rangle-CH_2-CH_2-\langle\bigcirc\rangle-$. The former begins to decompose at 320°, the latter at 470°. In the former the methylene group is linked directly to the imide ring, i.e., we have $\rangle N-CH_2-\langle\bigcirc\rangle-$, while in the latter the link is through a benzene ring, i.e., we have a group $\rangle N-\langle\bigcirc\rangle-CH_2-$. Thus, the N—CH$_2$ bonds are weaker than $\langle\bigcirc\rangle-CH_2$ bonds. It may be expected that all polyimides with N—CH$_2$ bonds in the main chain will have a low decomposition temperature.

The results /64/ of prolonged measurements of weight loss by samples of different polypyromellitimides at 325° in the air are shown in Table 13.

TABLE 13. Weight loss incurred by polypyromellitimides of structure

$$-N\langle\substack{CO \\ CO}\rangle\langle\bigcirc\rangle\langle\substack{CO \\ CO}\rangle N-\langle\bigcirc\rangle-R''-\langle\bigcirc\rangle-$$

at 325 C in the air /64/

Diamine	R''	Loss in weight, %			
		100 hrs	200 hrs	300 hrs	400 hrs
m-Phenylene diamine	—	3.3	4.3	5.0	5.6
Benzidine.	>—<	2.2	3.6	5.1	6.5
4,4'-Diaminodiphenyl oxide. . .	—O—	3.3	4.0	5.2	6.6
3,4'-Diaminodiphenyl oxide. . .	The same	3.4	3.8	5.1	7.2
4,4'-Diaminodiphenyl sulfide . .	—S—	4.8	5.8	6.8	7.9
3,4'-Diaminobenzanilide	—CO—NH—	2.0	4.2	6.2	9.8
3,3'-Diaminobenzanilide	The same	3.2	6.5	9.8	11.2
4,3'-Diaminobenzanilide	The same	4.3	7.8	10.8	11.9
4,4'-Diaminobenzanilide	The same	5.7	8.4	11.9	12.1
Isophthalyl-3-aminoanilide . . .	—NH—CO⌬CO—NH—	4.7	6.8	8.9	10.7
Isophthalyl-4-aminoanilide . . .	The same	6.9	9.4	14.4	20.4
4,4'-Methylenedianiline	—CH$_2$—	9.4	12.4	14.7	16.8
N,N-m-Phenylene-bis-4-amino-benzanilide	—CO—NH⌬NH—CO—	6.0	9.2	12.5	15.6
N,N-m-Phenylene-bis-3-amino-benzanilide	The same	6.2	8.3	14.0	20.3
4,4'-Isopropylidenedianiline . .	—C(CH$_3$)$_2$—	16.1	26.2	31.0	36.0

61

No. of polymer	Structure of repeating unit	Initial rate of weight loss, %/hour	Weight loss during 3 hours, %
I	(polyimide structure)	0.4	1.2
II	(polyimide structure)	0.45	1.4
III	(polyimide structure)	1.2	3.2
IV	(polyimide structure)	1.1	2.8
V	(polyimide structure)	0.5	1.6
VI	(polyimide structure)	1.1	1.3
VII	(polyimide structure)	3.0	6.2
VIII	(polyimide structure)	5.0	8.2
IX	(polyimide structure)	5.5	8.8
X	(polyimide structure)	6.0	11.2
XI	(polyimide structure)	1.6	4.6
XII	(polyimide structure)	6.0	12.6
XIII	(polyimide structure)	1.9	23.1
XIV	(polyimide structure)	8.0	15.3
XV	(polyimide structure)	25.0	31.2

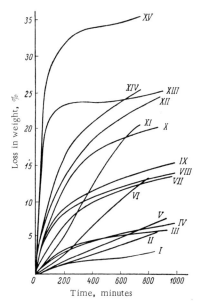

FIGURE 35. Loss of weight by polyimide samples of different structures at 450°C in an atmosphere of helium.

The numbers of curves correspond to the numbers of polymers listed in Table 14.

Of the polymers studied in /64/, numerous polyimides and polyamido-imides are technologically valuable products. Prolonged determinations of weight loss at relatively low temperatures are a suitable method for determining the thermostability of polyimides. It is seen, however, on comparing the data in Tables 11 and 13, that tests of brief duration, performed at elevated temperatures, will also give an idea of the thermal stabilities of polyimides.

In addition to polypyromellitimides, a large number of polyimides based on other tetracarboxylic acid dianhydrides have been recently prepared, but literature data on their thermostability and transformations produced by decomposition are practically nonexistent.

Some of our own data of the estimated thermostability (weight loss during brief periods of time) of a number of polyimides prepared from different anhydrides of tetracarboxylic acids are given below. The structural formulas of the polyimides are shown in Table 14. The final stage thermal treatment of the samples was performed in vacuo at 380° during one hour. The tests were conducted in helium at 450°. The curves of loss in weight as a function of time for all polymers are shown in Figure 35. The slopes of the curves usually decrease with time, especially so for polymers XIII and XV, which contain cyclopentane and pyridine rings in the dianhydride component; the rate of weight loss was highest for these polymers. The least loss in weight was noted for polymer I, which contains only ether groups in addition to aromatic and imide rings.

Table 14 shows the initial degradation rates, calculated from Figure 35 on the strength of weight loss during the first 60 minutes, and the total weight loss during 3 hours' heating.

On comparing the polyimides III, IV, V, IX, XIII, XV, based on diaminodiphenyl ether, the radical of which is especially highly stable, the radicals R of tetracarboxylic acid dianhydrides can be arranged in the following decreasing thermostability sequence:

63

This sequence also remains valid for polyimides prepared from other aromatic diamines.

It is seen from the data given in Table 14 for polyimides I, IV, VII and VIII, obtained from the dianhydride of bis-(3,4'-dicarboxyphenyl) oxide, that the R' radicals of the diamine form the following decreasing sequence of thermostability:

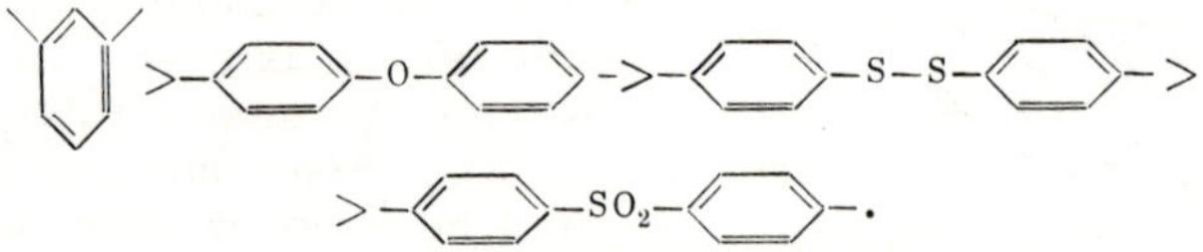

It will be noted that the thermostability of a polyimide markedly depends on the location of the given structural unit — in the dianhydride or in the diamine component of the link. For example, polyimides with the

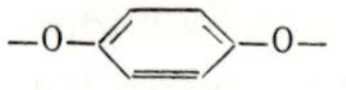

$-O-$⟨ ⟩$-O-$ group in radical R (polymers X, XI, XII) have a lower thermal

stability than those which contain this group in R' (polymers II and VI).

The results of the studies carried out on polyimides differing in both R and R' radicals make it possible to complement to some extent the conclusions of Bruck on the mechanism of degradation of polyimides.

If it is assumed, according to Bruck, that in an inert medium aromatic polyimides with phenyl nuclei interconnected by ether groups $-O-$ decompose solely by way of cleavage of the carbonyl group $C = O$ of the imide rings, the overall rate of this process may be expected to be a function of the size of the monomeric unit. Let us accordingly consider the series of polyimides represented in Table 15.

TABLE 15. Comparison of $C = O$ group concentration in various polyimides with their rate of weight loss

No. of polymer	Structure of monomer unit	M_0	$N_0 \cdot 10^{-21}$, g^{-1}	v, %/hour
XI	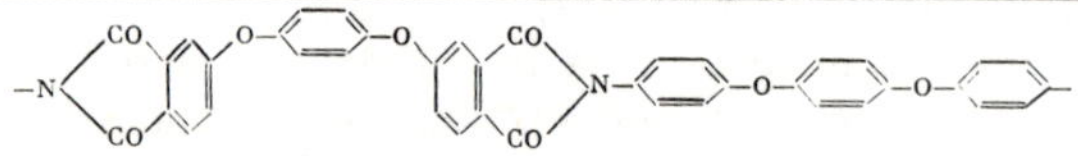	566	4.2	1.6
II	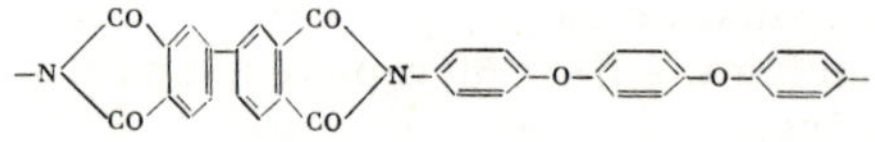	550	4.4	0.45
IV	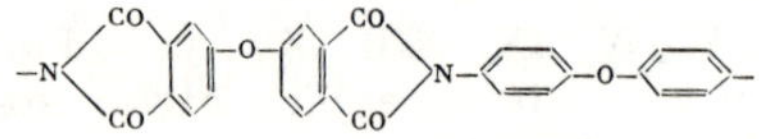	474	5.1	1.1
X	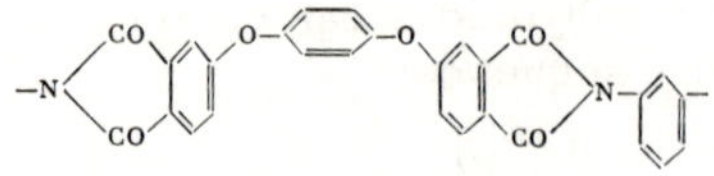	474	5.1	6.0
V	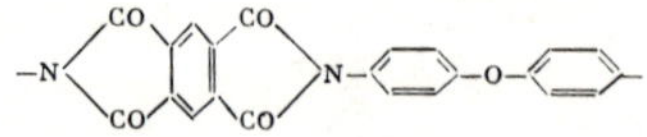	382	6.3	0.5
I	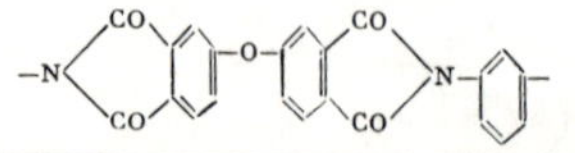	334	7.2	0.4

Note. The polymers are numbered as in Table 14. Values of v have been taken from Table 14.

64

These polyimides differ from one another only in the number or the sequence of phenyl nuclei and oxygen atoms. Since the molecular weights M_0 of the monomeric units are different, the compounds also differ in the concentration N_0 of carbonyl groups per unit weight. The polymers in Table 15 are arranged in increasing sequence of N_0. If it is assumed that the decomposition consists solely in the cleavage of $C = O$ groups in the imide ring, it could be expected that the rate of weight loss will increase with increasing N_0. In fact, as may be seen from the table, there is no connection at all between the two parameters. Polyimides with equal concentrations of $C = O$ groups (polymers II and XI, IV and X) may have weight loss rates differing by a factor of 5 —10.

This finding may be due to the fact that the rate of decomposition of imide rings depends to a large extent on the chain structure — for example, the activation energy of the $C = O$ group cleavage might not be a constant magnitude — or else that the mechanism of degradation of polyimide chains is different depending on the content and location of ether bonds in the monomeric unit, the presence of conjugated bonds between the imide rings, etc. No final conclusions on the subject can be arrived at on the strength of the available data.

It must finally be noted that degradation strongly affects the physical and mechanical properties of the polymers. As a rule, these properties begin to deteriorate long before the incipient decomposition can be detected by DTA or thermogravimetric methods. It follows that the decomposition temperatures as determined by these methods, especially if such determinations involve a continuous, rapid rise in temperature, give no indication of the service range of temperatures for the given polymer. The actual working temperature range under conditions of prolonged exploitation is always much lower than the decomposition temperature determined by DTA or by thermogravimetry.

Chemical stability of polyimides

Most polyimides, in particular the highly thermostable aromatic products of importance in technology, are inert to organic solvents and oils. They are also not significantly affected by dilute acids. Polyimides dissolve in strong acids, such as fuming nitric acid or concentrated sulfuric acid, especially in the heat /101/. The resulting solutions are unstable and their viscosity drops with time (Figures 36, 37). Films cast from these solutions are extremely brittle. It follows from these data that the dissolution is accompanied by degradation.

The resistance of polyimides to alkalis and superheated steam is low, and the polymers are hydrolyzed by these agents. Clearly, in the first instance, this is the result of the presence of $C = O$ groups in the ring.

Nishizaki /85/ made a thorough study of the mechanism of hydrolysis of polyimides, using polypyromellitimide

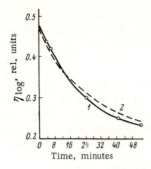

FIGURE 36. Logarithmic viscosity of solutions of polypyromellitimide in concentrated sulfuric acid at 30°C as a function of time /101/.

$1 - R' = -\langle\!\!\!\langle\ \rangle\!\!\!\rangle-CH_2-\langle\!\!\!\langle\ \rangle\!\!\!\rangle-$;

$2 - R' = $

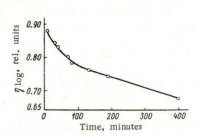

FIGURE 37. Logarithmic viscosity of a solution of polypyromellitimide

$\left(R' = -\langle\!\!\!\langle\ \rangle\!\!\!\rangle-O-\langle\!\!\!\langle\ \rangle\!\!\!\rangle-\right)$ in fuming

nitric acid at 15° C as a function of time /101/.

He treated polyimide films with solutions of alkalis and acids and recorded the resulting changes in the external aspect of the films, their solubility and IR spectra. The polyimide film (I) of 25μ thickness swells when immersed in 4N aqueous KOH at 20°, its color fades and in a few hours becomes colorless, while retaining sufficient strength and elasticity. The resulting film (II-K) is soluble in water, dilute (less than 1N) alkalis and aqueous ammonia of all concentrations, but not in concentrated alkalis, dimethylacetamide and other polar organic solvents.

The IR spectrum of film II-K (Figure 38, 3) has no typical imide bands at 1780, 1729 and 723 cm^{-1}, but amide bands — amide 1 (1682—1670 cm^{-1}) and amide 2 (1550 cm^{-1}) — are present. There is also a broad carboxylate absorption range /3/ at 1550—1600 and 1400 cm^{-1}. When treated with hydrochloric acid, II-K film contracts, becomes insoluble in water, but soluble in dimethylacetamide (film III). The IR spectrum of film III (Figure 38, 4) is very similar to that of the initial polyamido-acid film (Figure 38, 1) obtained directly by the reaction between pyromellitic dianhydride and diaminodiphenyl ether. The spectrum of film IV, obtained from film III by heating for 30 minutes at 140° (Figure 38, 5), again contains the imide bands at 1780, 1729 and 723 cm^{-1}. The film yellows slightly, becomes insoluble in dimethylacetamide and externally resembles film I. All this indicates that the polyimide is regenerated as a result of this sequence of operations.

The transformations of the polypyromellitimide films under these conditions may be represented by the following reaction scheme /85/:

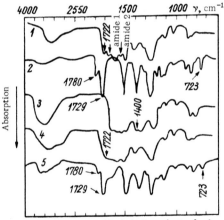

$$\left[-N \begin{array}{c} CO \\ \\ CO \end{array} \begin{array}{c} CO \\ \\ CO \end{array} N-\bigcirc-O-\bigcirc- \right]_n$$

(I) (IV)

↓ KOH

$$\left[\begin{array}{cc} KOOC & CO-NH-\bigcirc-O-\bigcirc- \\ -NH-OC & COOK \end{array} \right]_n$$

(II-K)

↓ HCl

$$\left[\begin{array}{cc} HOOC & CO-NH-\bigcirc-O-\bigcirc- \\ -NH-OC & CO-OH \end{array} \right]_{n.}$$

(III)

The regenerated polyimide film (IV) is more brittle and weaker than the initial film (I). It may be assumed /85/ that the alkali treatment results not only in the opening of imide rings, but also in the scission of the main chain. It should be remembered, however, that the temperature of thermal treatment of film III could have been too low for the imidization to proceed to completion. Thus, the difference between the properties of the initial and the regenerated polyimide films could have also been largely due to the differences in the degree of imidization.

FIGURE 38. IR spectra of hydrolysis products of polyimide

$$\left[-N \begin{array}{c} CO \\ \\ CO \end{array} \begin{array}{c} CO \\ \\ CO \end{array} N-\bigcirc-O-\bigcirc- \right]_n \quad /85/.$$

1 — polyamido-acid; 2 — polyimide (film I); 3 — film II-K, product of hydrolysis of polyimide by KOH; 4 — film III, obtained by treating film II-K with hydrochloric acid; 5 — film IV, obtained by heating film III at 140°C for 30 minutes (regenerated polyimide).

67

TABLE 16. Elementary analysis of products of treatment of polypyromellitimide
with different alkalis /85/

Sample	Content of alkali metal, %	
	calculated	found
II-K	15.81	14.4
II-Na	9.94	10.02
II-Li	3.22	3.89

Similar transformations also take place when the polyimide film is treated with other alkalis — NaOH and LiOH. Results of elementary analysis show that the main products of hydrolysis are alkali salts of the polyamido-acid (Table 16), in accordance with the scheme given above.

TABLE 17. Rate constants of decoloration of polyimide film
by solutions of KOH /85/

Normal concentration of aqueous KOH	Rate constant k, min^{-1}
5	0.0500
3	0.0133
2	0.0059
1	0.0029

As has already been pointed out, the polyimide film loses color as a result of hydrolysis. This effect was taken advantage of /85/ to determine the rate constant of hydrolysis under different conditions. Figure 39 shows the absorption of visible light as a function of the wavelength for films held in 3N solutions of alkali for different periods of time. If the duration of hydrolysis is increased, the curves shift to the left. The optical density D_{460} at 460 mμ was taken as the index of the degree of transformation of the polyimide into alkali salt of the polyamido-acid. This index was utilized to calculate the rate constants of the reaction at different concentrations of the alkaline solution. Figure 40 shows the variation of the optical density D_{460} of the films as a function of the time t at different alkali concentrations. In a wide range of D_{460} values this function is linear:

$$D_{460} = -kt + C.$$

Table 17 shows the rate constants k of decolorization for different concentrations of aqueous solution of KOH, calculated from Figure 40.

As regards their hydrolytic activity, the alkalis can be arranged in the sequence: KOH > NaOH > LiOH. Polyimide films also become decolorized in aqueous 2.5N solution of ammonia, but at a slower rate than in solution of NaOH of the same concentration. The film dissolves at the same time.

Saturated barium hydroxide is practically without effect on polyimides at room temperature. At 80°, the film becomes fully decolorized in 3 hours and becomes brittle and insoluble in water and in dimethylacetamide.

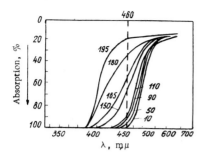

FIGURE 39. Variation of the absorption of visible light during the treatment of polyimide film

with 3N aqueous KOH at 21° C /85/.

Numbers marking the curves indicate the time of hydrolysis in minutes.

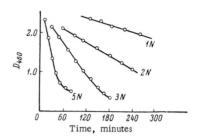

FIGURE 40. Decolorization of polyimide film

during hydrolysis with aqueous solutions of KOH of different concentrations /85/.

D_{460} — optical density at 460 mμ.

In saturated calcium hydroxide solution there is no decolorization and no formation of the calcium salt of the polymer.

TABLE 18. Stability of polypyromellitimide films on being boiled in water /101/

Diamine radical R'	Duration of boiling during which flexibility preserved
	1 week
	2 weeks
	1 year
	> 3 months

In aqueous solutions of metal salts, type II-K film is capable of undergoing cation exchange. Thus, when immersed in a solution of ferrous chloride it darkens, becomes hard and brittle and insoluble in water and in dimethylacetamide. Analysis showed that such a film contains 7.83% iron and less than 0.5% of potassium. The iron ions can be eluted by treating the film with hydrochloric acid, when the polyamido-acid is again formed. Similar effects have also been noted for salts of several other metals — lead, copper, nickel, mercury, manganese and chromium. The cation exchange is accompanied by different colorations of the film.

Pure water is without effect on polypyromellitimide films under ordinary conditions. However, when boiled in water, the films lose their superior mechanical properties. Table 18 shows comparative data on the water resistance of a number of polypyromellitimides under these conditions.

It is seen from the data in Table 18 that polypyromellitimide based on 4,4'-diaminodiphenyl ether $\left(R' = - \langle \underline{} \rangle - O - \langle \underline{} \rangle - \right)$ is the most water-resistant.

When heated in a sealed ampoule containing some water, for 300 hours at 160°, this film, while not decomposed, became somewhat decolorized. The 3620 cm^{-1} hydroxyl absorption band in its IR spectrum became markedly intensified. Under similar conditions the polyester film "Mylar" and poly (vinyl fluoride) film "Tedlar" became brittle or decomposed altogether /92/.

Chapter III

PHYSICAL PROPERTIES OF POLYIMIDES OF DIFFERENT CHEMICAL STRUCTURES

The high reactivity of the dianhydrides of tetracarboxylic acids and diamines makes it possible to introduce considerable variations in the structure of the resulting polyimides and to study the relationships govern-ing the consequent variations in their physical properties. A number of such results will be found in the publications /20, 30/ by the authors of this book. In what follows we shall give a detailed description of the physical properties and of the experimentally established correlations with the chemical structures of various polyimides and copolyimides prepared by the two-step method of synthesis, in which the second step — the imidization of polyamido-acids — was effected by curing at temperatures up to 300—400°. The structure of radicals R and R' in the dianhydrides and diamines used is represented in Table 19. For the sake of convenience, we shall denote the chemical structure of polyimides, both in the text and in the captions to figures, by a combination of Roman (dianhydride radical) and Arabic (diamine radical) numerals in accordance with Table 19. The most frequently mentioned polymers will also be denoted by the letter symbols shown in Table 19.

The presence in the polymer chains of benzene rings interlinked by means of five-membered imide rings and heteroatom groupings is responsible for the numerous properties which they all have in common. On the other hand, the variation in the locations of the benzene rings relative to the imide rings and hetero-atomic groups gives rise to characteristic differences and makes it possible to propose a structure vs. properties classification of the entire series of polyimides. We shall begin by discussing the properties which these polymers have in common.

Common properties and characteristic differences of polyimides

Polyamido-acid films are faintly colored, transparent, often very strong and elastic.* For example, the film of polyamido-acid PM(I-5) at

---

* Here and in what follows elasticity will be understood to mean the capacity of the polymer to undergo considerable deformation without disintegration. The measure of elasticity is taken to be the elongation-at-break. The data for the physical and mechanical properties of polyimides given in this chapter were mostly obtained by testing films $20-50\,\mu$ thick. Measurements of static elasticity modulus and all deformation strength and thermomechanical parameters were conducted with the aid of the universal instrument UMIV-3. Dynamic measurements were performed on a special installation, operating in $20-200$ hz frequency range and $20-600°$ temperature range.

TABLE 19. Structure of polyimides discussed in Chapter III.
A. Radicals of dianhydrides R and diamines R'

R		R'	
No.	Formula	No.	Formula
I		1	
		2	
II		3	
III		4	—⟨ ⟩—SO₂—⟨ ⟩—
		5	—⟨ ⟩—O—⟨ ⟩—
IV	(CO)	6	
V	(SO₂)	7	
		8	
VI	(O)	9	
		10	
VII		11	
VIII		12	
IX		13	

TABLE 19 (continued)
B. Arbitrary designations of certain polyimides

Structure of polymer	Designation of polymer by nos.	by letters
	I-5	PM
	III-9	DF-FG
	VI-3	DFM
	VI-5	DFO
	I-9	PFG

72

20° has a tensile strength of about 1200 kg/cm² and an elongation-at-break of about 80%. Polyamido-acid films are, however, very unstable. On being stored for several months in the air at room temperature they become very brittle owing to hydrolysis by atmospheric moisture. Polyimide films obtained by heating such partly hydrolyzed polyacid films are very strong and elastic. Polyacid films stored in a dry atmosphere display less extensive deterioration in properties.

Polyimide films are transparent, and have a more or less intensive yellow coloration. On being heated at 300−400° they turn reddish-brown. The films of polyimides VI-4, VII-4, VII-9 and other low-conjugated polymers are almost colorless. The coloration often depends on the degree of purification of the starting substances. For example, carefully purified 4,4'-diaminodiphenyl ether gives a light-yellow PM film, while the unpurified compound yields a reddish-brown film.

Reactions of polyimide formation in the solid phase by heating the corresponding aromatic polyamido-acids proceed at a high rate in the same temperature range. This may be seen from the measurements of the tangent of dielectric loss angle (Figure 41), density and refractive index (see below, Figures 84, 90), static and dynamic elasticity modulus (see below, Figures 46, 83) and other physical characteristics of aromatic polyamido-acids of different chemical structures during the thermal treatment. Imidization of polyamido-acids based on aliphatic and alicyclic dianhydrides proceeds in a wider temperature range and at somewhat slower rates, possibly owing to the weakening of the system of conjugated bonds (cf. Figure 17).

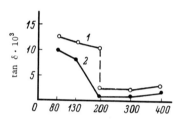

FIGURE 41. Tangent of dielectric loss angle (20°C, 5 · 10⁴ hz) as a function of the heating temperature of the initial polyamido-acid films for polymers I-5 (curve 1) and III-2 (curve 2).

Time of heating 15 minutes at each temperature. Times of heating at 200°C for curve 1 were 5 and 20 minutes.

In all cases thermal imidization must be performed in thin layers in order to prevent hydrolytic decomposition of high polymers. Thermal treatment of polyamido-acids of polyimides with a softening point (e.g. VI-5) in thick layers or at a high heatup rate is accompanied by blistering. Nonsoftening polyimides (e.g., I-5) disintegrate into powder under these conditions, while polymers based on the dianhydride of 2, 3, 4, 5-cyclopentanetetracarboxylic acid (e.g., VIII-5 and VIII-9) form fairly strong foams recalling plastic foams in appearance.

Thermal treatment in thin layers may be performed in most cases under standard conditions. In the simplest case — preparation of polyimide films — the solution of the polyamido-acid is poured onto a support (glass, metal, etc.), dried at 20-100° to produce a film, and the film is then heated at 130, 150, 200 and 250° for 15−30 minutes each time. The treatment is terminated by curing at 300° for one hour in an inert medium. In certain cases, e. g., polymer PM (I-5), the properties of the film are improved by curing at 400° for 15 minutes in an inert gas atmosphere. Films dried on supports have mechanical properties superior to those of films which have been heated free. The reason for it is the spontaneous biaxial orientation, which is the result of shrinkage stresses appearing during imidization on supports. The strength and the elasticity modulus of polyimides may also be considerably enhanced by subjecting the polyamido-acids to a preliminary orientation stretching. This indicates that cyclodehydration is not accompanied by marked conformational changes in the macromolecules.

Insolubility in organic solvents is displayed by most aromatic polyimides. Polyimides such as PM (I-5) do not dissolve and do not swell even on prolonged heating. Polymers which soften, such as VI-5, IX-5 or IX-9, swell in hot dimethylformamide. The introduction of side groups into the chain, such as CH_3 groups, results in low-density polyimides (see below, Table 21). Such polymers, which have a loosened structure (e.g., IV-12), may completely dissolve in dimethylformamide at 20°. However, if polyimide film VI-12 is heated above 300°, an insoluble fraction appears, and its amount becomes larger as the time and temperature of the treatment are increased.

The location of R and R' radicals in the polyimide chain is perfectly regular. This is ensured at the polyamido-acid preparation stage. The main imidization reaction is accompanied mainly by the formation of linear, nonbranched, somewhat intertwined polymer chains. This is confirmed by the high fluidity of the softening polyimides (VI-5, VI-7 etc.), by the fact that a large number of them crystallize, and by the fact that the soluble polyimides dissolve without previous swelling, which is typical of linear systems. The disappearance of fluidity, the appearance of the insoluble fraction and other effects noted after the high-temperature treatment are the result of secondary chemical reactions which result in crosslinking. Their effectiveness will depend on the chemical structure of the polyimide.

A number of properties of polyimides in the bulk are determined by the strong intermolecular forces connected with the presence of benzene rings and imide rings in the chains. Thus, polyimides typically have high temperatures of phase transitions. Their softening temperatures are usually above 200° and their melting temperatures above 400° (see Table 24 below). Isotropic polyimide films have high moduli of elasticity. At room temperature their values are between $30 \cdot 10^3$ and $100 \cdot 10^3$ kg/cm^2 (Table 24). The elasticity moduli of carbon-chain polymers in the vitreous state are much lower — between $20 \cdot 10^3$ and $30 \cdot 10^3$ kg/cm^2. The elasticity modulus increases even more as a result of orientation stretching. Thus, when PM (I-5) films are thermally stretched by 150%, the static elasticity modulus at 20° increases from 35,000 to 80,000 kg/cm^2; after stretching by 100% at room temperature the modulus becomes twice as large. The dynamic elasticity modulus of an isotopic film of DFO (VI-5) at 30 hz and 20° is 32,000 kg/cm^2; after 100% stretching it increases to 60,000 kg/cm^2. Elasticity moduli show even larger increases as a result of fiber stretching. The increase of elasticity modulus as a result of stretching is a typical effect given by crystalline polymers. Amorphous polymers as a rule do not display such effects to any significant extent. In the case of polyimides, however, this effect is displayed even by polymers such as DFO which are obviously amorphous. Polymer PM (I-5), which can crystallize /101/, on being thermally stretched displays a marked increase in elasticity modulus, but its density remains unchanged. This would seem to mean that here, too, a marked effect of orientation crystallization is absent. When the orientation is accompanied by a strong increase in the density and crystallization, the elasticity modulus may become as high as 300,000 — 400,000 kg/cm^2, i.e., may assume values typical of the elasticity moduli of inorganic glasses and metals.

The strong intermolecular forces are responsible for the high dimensional stability — small deformation rates under constant loads — of aromatic polyimides at high temperatures. Thus, for instance, polyimide DFO (VI-5), with a softening point of 265°, has a creep rate at 200° which is 1000 times lower than that of polypropylene at 23° under the same constant tensile strength (Table 20).

TABLE 20. Deformation of polymers under constant stretch loads

Temperature, °C	Load, kg/cm^2	Creep rate, % · 10^{-3}/hour			
		polypropylene /91/	polybutene /91/	polyphenylene oxide /63/	polyimide DFO
23	120	50	1.5	—	0.7
	140	1000	2.0	2.5	1.0
	180	—	—	—	1.7
200	120	—	—	—	0.8
	140	—	—	—	1.2
	800	—	—	—	400

It is seen from Table 20 that the creep rate of DFO does not significantly vary with the temperature. This property is of major practical importance.

The density of polyimides, which is high for polymers, is due to a strong intermolecular interaction. It will be seen from Table 21 that the density values vary between 1.28 and 1.48 g/cm^3. The highest densities are displayed by polymers with rigid small aromatic monomeric links. If the size of the monomeric link is increased and hetero-atoms or side groups introduced, the density decreases. Some polymers acquire the capacity to crystallize as a result (densities in the crystalline state are shown between brackets in Table 21). The lowest densities are displayed by polymers with supplementary side groups in the main chain (I-12, VI-12). Polymers with disulfide grouping —S—S— are always much denser than those with the sulfide group —S— and otherwise identical chemical structure (I-7 and I-8, III-7 and III-8, VI-7 and VI-8, Table 21). The softening temperatures of polymers with disulfide groups are lower.

TABLE 21. Densities at 20°C of polyimides of different structures

Polymer	Density, g/cm^3	Polymer	Density, g/cm^3	Polymer	Density, g/cm^3
I-2	1.455	III-7	1.370	VI-7	1.378
I-5	1.420	III-8	1.399	VI-8	1.406
I-7	1.402	III-9	1.351 (1.382)	VI-9	1.350 (1.380)
I-8	1.477	III-10	1.352 (1.397)	VI-10	1.355
I-9	1.400	IV-5	1.379	VI-12	1.280
I-10	1.390	V-4	1.450 (1.460)	VI-13	1.426
I-12	1.276	V-5	1.420	VII-9	1.325 (1.395)
II-5	1.417	VI-1	1.477	VIII-5	1.380
II-7	1.410	VI-2	1.420	VIII-9	1.340
II-9	1.408	VI-3	1.407 (1.427)	IX-1	1.400
III-1	1.465	VI-4	1.416	IX-2	1.380
III-2	1.410	VI-5	1.380	IX-5	1.360
III-5	1.408	VI-6	1.380	IX-9	1.340

The densities of polyamido-acids are always much lower than those of polyimides. For example, in the case of polymer I-2 the densities of polyamido-acid and polyimide are 1.348 and 1.455 g/cm^3 respectively, while those of polymer VI-2 are 1.345 and 1.420 g/cm^3 respectively. Thus, imidization is always accompanied by contraction.

The density of polyimides will depend on the conditions of imidization. Thin films, out of which the solvent can readily diffuse during the thermal treatment, have lower densities. For example, the density of polyimide I-5 varies, depending on the film thickness, between 1.410 for thin films ($2-10\mu$) up to 1.428 g/cm^3 for films 100μ thick. It is indicated by IR spectra and by studies of mechanical properties that imidization in thin films takes place at higher temperatures. This seems to be accompanied by the fixation of the less dense, high-temperature packing of the molecular chains. The presence of the solvent, which is unavoidable in thick layers, reduces the imidization temperature and facilitates crystallization.

Formation of stable free radicals at high temperatures is a typical feature of polyimides. The radicals are readily detected by EPR/6, 45, 49/. Radicals formed on heating above 200° (singlet, $\Delta H = 6-8$ oersteds) are the most typical. They are encountered in concentrations of $n = 10^{17} - 10^{19}$ spins/cm^3, and at 20° in nonsoftening polyimides recombine very slowly. For example, in the polymer I-5 their concentration decreased by only $10-20\%$ during one month. The residual free radical concentration in softening polyimides, e.g. in VI-5, is much lower than in those which do not soften. The high stability of the radicals to oxidation and other effects permit to relate them to the nitrogen RPP atom. They may be formed (cf. pp. 36, 42) as a result of disruption of imide rings followed by incomplete recombination.

As regards their mechanical strength at room temperature, there are no marked differences between polyimides of different structures if their molecular weights are sufficiently high. On the other hand, the deformation capacity varies strongly with the structure. Polypyromellitimides with hetero-atoms in R', and aromatic polyimides with hetero-atoms in R and R', can be subjected to major forced elastic deformations. For example, polymers I-5, I-9, VI-5 and VII-9 have a 100% elongation-at-break at 20°. Polypyromellitimides I-5 and I-9 retain satisfactory elasticity at −195°, while polymer VI-5 retains it down to −150°. Polyimides without "hinge" hetero-atoms in R' have a low elasticity. Thus, the polymer based on pyromellitic acid dianhydride and benzidine (I-2) at 20° has an elongation-at-break as low as $1-2\%$. A decrease in the molecular weight of the polyamido-acid brings about a decrease in elasticity and mechanical strength if the polycondensation is difficult, e.g. in polymers VI-4 and V-4. Elasticity, especially at low temperatures, is often closely connected with the relationships governing secondary chemical transformations which take place during the thermal treatment. The nature of the transformations and their effect on the mechanical properties will depend on the chain structure.

Very high thermal and thermooxidative stability is displayed by aromatic polyimides. The introduction of aliphatic groups $-CH_2-$, $-C(CH_3)_2-$ etc. reduces the stability. This means that the stability is determined by those parts in the chain which are least resistant to heat. Low thermostability is displayed by polyimides based on dianhydrides of pyridinetetracarboxylic acid and aliphatic and alicyclic tetracarboxylic acids. The weak spots in the chain are the dianhydride radicals. In fully aromatic polyimides the thermal stability is usually determined by the stability of the imide ring conjugated with the phenyl nuclei. These problems have been discussed in detail in the preceding chapter. It should be added, however, that owing to the high thermal stability of the macromolecules of aromatic polyimides, the physical properties of these polymers remain very stable under prolonged exposure to heat. For example, polymer PM (I-5) in the form of a film retains the minimum permissible mechanical parameters at 350° for one year in an inert atmosphere and at 250° for 8—10 years in the air ("H-film," see following chapter). The mechanical strength of DFO polymer decreases by not more than 10% as a result of 500 hours' holding in the air at 250°. Carbon-chain polymers show a marked deterioration in properties within shorter periods of time at lower temperatures. For example, nonstabilized polypropylene loses 90% of its strength during 115 hours at 125° in the air, whereas the polymer PM displays similar losses in strength during similar periods of time only at 400°. This can be seen in Table 22, which contains the results of thermostability tests carried out on some aromatic polyimides by determining the variation in their mechanical properties.

TABLE 22. Changes in mechanical properties of polyarimides as a result of heating in helium at high temperatures

Polymer	Tensile strength, kg/cm^2			Elongation-at-break, %		
	initial sample	after 3 hours holding at		initial sample	after 3 hours holding at	
		400°	450°		400°	450°
I-5	1800	1700	1200	70	50	9
IV-5	1600	1580	1400	16	15	7
VI-5	2000	1600	1000	90	26.5	4.5

All polyarimides shown in Table 22 have about equal values of thermal stability, as estimated from the loss in weight (Figure 35), but the mechanical properties of each vary at different rates as a result of aging. This indicates that the thermostability of physical properties of polymers can be measured by the results of gravimetric analysis only very approximately.

The stability to ionizing radiation of aromatic polyimides is very high. It will be shown below (Chapter IV) that they withstand very large doses of γ-radiation, neutrons, electrons and UV light. This property is obviously due to the chain-ring structure of polyimides.

As regards their electrical properties, polyimides are medium-frequency dielectrics. Their dielectric constant ϵ is 3—3.5 and does not change much with the frequency and with the temperature. The specific volume resistance ρ_v at room temperature is 10^{17} to 10^{18} ohm · cm, while at 200° it is about 10^{14} ohm · cm. Above 200° the temperature variation of specific resistance is determined by the softening temperature. The tangent of the dielectric loss angle $\tan \delta$ at room temperature remains practically constant as the structure of the chain varies (Table 23). In all cases $\tan \delta = 1 \cdot 10^{-3} - 1.5 \cdot 10^{-3}$; polyimides with a disulfide grouping in the chain (VI-8 in Table 23), for which $\tan \delta \simeq 0.6 \cdot 10^{-3}$, are an exception. These polymers have a high density (see above, Table 21). It would appear that in the case of polyimides dielectric losses are caused mainly by the polar CO groups in the imide ring. The symmetrical location of these groups is responsible for the low loss level — the tangent of the loss angle is 2—4 times lower than in polyesters poly(ethylen terephthalate) in which the CO groups form part of the main chain.

TABLE 23. Dielectric losses by a number of polyimides at 23°C and a frequency of $5 \cdot 10^5$ hz

Polymer	$\tan \delta \cdot 10^3$	Polymer	$\tan \delta \cdot 10^3$
I-2	1.1	IV-5	0.9
I-5	1.2	VI-8	0.6
II-7	1.6	VI-3	1.0
III-2	1.0	VII-3	1.5

Detailed temperature-frequency relationships of dielectric parameters (ϵ and tan δ) were determined only for a commercial polyimide — "H-film" /38/. It has been shown /47/ that this product consists of polyimide based on pyromellitic acid dianhydride and 4,4'-diaminodiphenyl ether (polyimide PM, I-5). The dielectric parameters of the "H-film" have been determined /38/ in the 100—100,000 hz frequency range and between — 70 and +230°. It is seen from Figure 42 that ϵ and tan δ do not markedly vary with frequency and with the temperature between 50 and 230°. Between + 40 and —70° the dielectric constant and the tangent of dielectric loss angle increase and become a function of the frequency. There appears a maximum of dielectric losses, which shifts towards higher temperatures as the frequency is increased. There is also a small tan δ maximum at elevated temperatures (about 150°). The location of this maximum is practically independent of the frequency. The same maxima are also found on the curve of temperature dependence of the tangent of loss angle for pure polypyromellitimide PM (Figure 42b).

The reason for the high-temperature maximum of the losses is unkown. We can merely note that it is located in the softening range of the polyamido-acid corresponding to polymer PM (see Figure 88 below), and may be connected with this process.

The magnitude of low-temperature dielectric losses, the frequency shift of their maximum and the temperature range in which they are noted are very similar to the corresponding magnitudes of low-temperature losses in poly(ethylene terephthalate), for which it has been conclusively demonstrated /95/ that the low-temperature losses are caused by the presence of terminal hydroxyl groups.

If the frequency f_m, corresponding to maximum dielectric losses as given by Figure 42b, is plotted as a function of reciprocal temperature (Figure 43), we may find the activation energy of the low-temperature relaxation process from the formula

$$f_m = A e^{-\frac{U}{RT}}.$$

78

The activation energy U for the "H-film" is 9.5 kcal/mole (the value of 15.4 kcal/mole given in /38/ is erroneous). For poly(ethylene terephthalate) $U = 12.4$ kcal/mole, which is fairly close to the U-value of the polyimide.

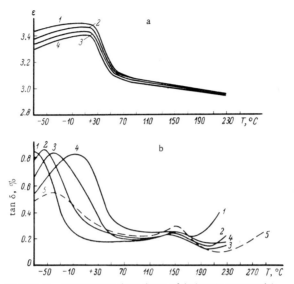

FIGURE 42. Temperature dependence of dielectric constant (a) and tangent of dielectric loss angle (b) for the "H-film" /38/.

Frequencies (hz): 1 — 100, 2— 1000, 3— 10,000, 4— 100,000. Broken line (b) — data /19/ for pure polypromellitimide at 1000 hz.

It has been suggested /38/ that the reason for the low-temperature losses in the polyimide might be the effect of the terminal COOH- and NH$_2$-groups, but confirmatory tests of the kind conducted in /95/ are required for a final conclusion.

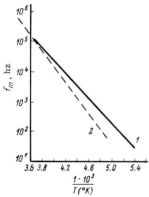

FIGURE 43. Frequency of maximum dielectric losses as a function of reciprocal absolute temperature for "H-film" (1) and poly(ethylene terephthalate) (2) /38, 95/.

The low dielectric losses of the "H-film," as in the case of other polyarimides, are clearly due to the fact that the bulk of the polar carbonyl groups is found in symmetrical ring groupings.

The increase in the dielectric losses in polypyromellitimide at 250° and above (Figure 42b and /19/) is mostly caused by conductivity losses. In general, dielectric values (see also Chapter IV) in conjunction with thermomechanical data /36/ indicate that polypyromellitimides do not dispaly distinct relaxational processes due to softening and vitrification in a narrow temperature range.

Classification of aromatic polyimides according to their physical and mechanical properties and structure

Even though all aromatic polyimides display a number of properties in common, which are a direct consequence of the presence of aromatic and heterocyclic groupings in the chain, many such polymers, including those of very similar chemical structure, may show major differences as regards their physical and mechanical properties and phase transitions.

Thus, many polyarimides do not melt or soften below their decomposition temperatures. Others become highly elastic in a very narrow range of temperatures. Some are brittle both at room and at elevated temperatures; others remain flexible at cryogenic temperatures, down to the temperature of liquid helium. We shall now give a few examples.

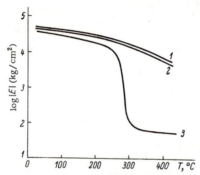

FIGURE 44. Dependence of the absolute dynamic elasticity modulus $|E| = \sqrt{E'^2 + E''^2}$ on the temperature for polyimides I-5(1), I-9 (2) and VI-5 (3). Frequency 30 hz.

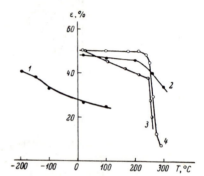

FIGURE 45. Longitudinal relaxation of oriented films of polyimides I-5(1, 2) and VI-5 (3,4) as a result of heating.

Extent and temperature of stretching:
1 — 40%, —195°; 2 — 48%, 270°;
3 — 50%, 20°; 4 — 50%, 250°.

Figure 44 shows the temperature dependence of the absolute dynamic elasticity modulus

$$|E| = \sqrt{E'^2 + E''^2}$$

for the polyarimides PM (I-5), PFG (I-9) and DFO (VI-5). It is seen that DFO has a narrow softening range and is highly elastic above 270°. Polymers PM and PFG (that latter has as many ring groupings and oxygen atoms in the monomeric unit as DFO) do not soften.

At room temperature PM and DFO have equal strengths and elasticities, but DFO is not elastic below —150°, while PM has an elongation-at-break of 30—40% or more even at —195°.

It is seen from Figure 45 that oriented DFO films strongly contract on heating near the softening point only, irrespective of their stretching temperature. On the other hand, longitudinal relaxation of oriented PM samples may occur at different temperatures: films stretched at —195° relax on attaining room temperature, those stretched at 20° relax at

200 — 250°, etc. If, however, PM films have been stretched at 400°, their ability to contract on heating is lost.

If we compare the physical and mechanical properties of aromatic polyimides with the chemical structure of the monomeric unit, we note certain general relationships between the structure and properties and may subdivide the polymers into a number of large groups. We shall take as the criterion for each group the presence in the monomeric unit of "hinge" links and their location with respect to the imide rings [20, 30]. Such "hinge" links may consist of hetero-atoms and hetero-groups $-O-$, $-S-$, $-S-S-$, $= SO_2$, or $= CO$, $-CH_2-$, $-C(CH_3)_2-$ or of benzene rings with bonds in the meta-position. These "hinges" render the molecule more flexible and thus facilitate its conformational rearrangements.

If this classification is adopted, each group will include polyimides which have not only a common chemical structure but also similar major physical and mechanical parameters.

Table 24 shows the deformational and mechanical strength parameters and temperatures of physical transitions (softening point T_s and melting point T_f) for a number of polyimides, arranged in four groups in accordance with the principle just stated.

Group A polyimides contain only aromatic rings, linked to each other directly or via imide rings. These are strong polymers, which have elasticity moduli of the order of $100 \cdot 10^3$ kg/cm^2 at 20°, brittle, nonsoftening. The elasticity modulus of polymer I-2 at 400° is more than $30 \cdot 10^3$ kg/cm^2, i.e., higher than in most common carbon-chain polymers at 20°

Group B polyimides contain hetero-atoms in the dianhydride component only. These hetero-atoms join together the benzene rings rigidly connected through imide rings to the aryl radical of the diamine component. It would appear that the presence of the hetero-atom is insufficient to ensure the rotation of the bulky rigid ring groupings around the bonds connecting them to the hetero-atom. Thus, in polymer VI-2 such a grouping consists of the structure

which is known to be flat. As a result, Group B polyimides are also nonmelting, nonsoftening and their elasticity is low.

Group C polyimides contain hetero-atoms in the diamino component only. These polymers give strong, resistant elastic films. The elongation-at-break — say, for polymer PM (I-5) — is 40—50%, even at —195°. Group C polyimides clearly display cross-linking. Their elasticity modulus increases with the time of residence at high temperatures. If the sample is rapidly heated to 400°, the elasticity modulus is of the order of 100 kg/cm^2 during the first few seconds and rapidly rises to $10^3 - 10^4$ kg/cm^2 (Table 24). The "hinge" groupings in polymer chains of this class are located between the phenyl rings which are in turn bound to the imide ring nitrogen by a single bond. This form of the molecule facilitates intramolecular rotation and ensures a looser packing density. Thus, for example, the density of polymer VI-2 (Group B) is 1.420. When the hetero-atom is in the diamine component (polymer III-5, Group C), the density decreases to 1.408. Group C polyimides have no definite narrow softening temperature range (Figures 44, 45, polymers I-5 and I-9).

81

TABLE 24. Properties of a number of representatives of four groups of aromatic

Group	R		R'	
	formula	No.	formula	No.
A		I		2
		II		1
		III		2
B		VI		1
		VII		2
C		I		5
		I		9
		II		9
		III		9
D		VI		5
		VI		9
		VI		4
		VI		3
		VII		4
		VII		9

* Amorphous films.
** At 300°.
† Maximum values.
†† Samples subjected to preliminary heat treatment (up to 270°).

polyimides

Tensile strength, † kg/cm²			Elongation-at- break, † %			Temperatures, °C		Elasticity modulus †† kg/cm²		
−195°	+20°	+400°	−195°	+20°	+400°	of soften- ing T_s	of fusion T_f	20°	400° 3 min	400° 15 min
—	2000	1000	—	2	15			120000	30000	33000
—	1500	—	—	4	—			75000	9000	—
—	2300	—	—	8	—	Do not soften		70000	8000	13000
—	1500	—	—	5	—			65000	9000	13000
—	1500	—	—	4	—			39000	5000	8000
500	2000	400	40	100	120			35000	500	6000
000	2000	400	30	130	140			32000	400	5000
—	1500	—	—	50	—	Soften, but become rap- idly cross- linked.		27000	200	2500
—	1800	250	—	20	160			35000	100	2000
2500	2000	—	8	100	—	270	—	30000	20	100
—	1500 *	—	—	50 *	—	250	450	27000	—	—
—	—	—	—	—	—	300	—	30000	45	45
—	1500 *	—	—	30 *	—	~260	490	29000	50 **	—
—	1100	—	—	15	—	290	—	32000	6	10
—	1500	—	—	100	—	200	400	24000	50	200

Group D polyimides contain "hinge" groupings in both the dianhydride and the diamine components. These polymers are elastic and have
the lowest density. Their most typical feature is the narrow softening
temperature range and transition to a visco-fluid state. Many of them are
capable of crystallizing, and have sharp softening temperatures and melting
points of the crystalline phase. The elasticity moduli at temperatures
above T_s are $10-100 \, kg/cm^2$, which is typical of the highly elastic state.
The cross-linking effects at elevated temperatures are weak. For example,
the elasticity modulus of polymer VI-4 (Group D) remains practically
unchanged after 15 minutes holding at 400°, whereas in the case of Group C
polyimides it increases 100 times or more under these conditions.

It may be concluded on the strength of numerous experimental data that
the properties of a polyimide are determined by the location of the "hinge"
hetero-atoms relative to the imide rings. Thus, polymers VI-1 and I-5
have an equal number of phenyl nuclei and one ether bond each. The former
(Group B) polymer is brittle, whereas the latter (Group C) is very highly
elastic (Table 24). Polymers I-9 (Group C) and VI-5 (Group D) each contain
four phenyl rings and two hetero-atoms. The former softens in a very wide
temperature range, has an elasticity modulus of 5000 kg/cm² at 400°, whereas the latter softens within a narrow temperature range and above 270°
passes into a highly elastic state with a low modulus of elasticity (Figure 44).
At the temperature of liquid nitrogen polymer I-9 is elastic, whereas
polymer VI-5 is brittle.

While comparing the data of Table 24 for polyimides in Groups B and C
it may also be noted that the hetero-atom in the diamine component imparts
to the polymers a much higher elasticity than a hetero-atom in the dianhydride component (see, for example, VI-1 and I-5).

Secondary transformations and physical properties

An important reason for the differences in properties between the
various aromatic polyimides are the special features of secondary processes
which take place after the completion of the imidization reaction proper,
and which are in turn connected with the link structure. In polymers
belonging to different groups these processes take place differently, at
different rates and often affect the physical properties of the polymers
in an opposite manner. The main result of these processes, which have
a highly specific effect on Group C polyimides is the formation of intermolecular bonds (cross-linking).

The course of cross-linking is most simply followed by following
the changes in the static or dynamic elasticity modulus with time at
elevated temperatures. Figure 18 above represents an example of such
determinations for the case of polymer I-5. It has been seen that the
static elasticity modulus shows sharp changes not only as a result of
imidization, but also after its completion, if the experimental temperature
is sufficiently high.

Figure 46 shows a diagram illustrating the effect of temperature on the
relative temporal changes of the static elasticity modulus in a wider
temperature range. Films made of PM (I-5) polyamido-acid, 30μ in
thickness, were preliminarily dried at 40°. The elasticity modulus I

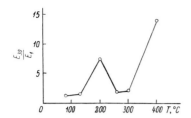

FIGURE 46. Relative change of static elasticity modulus with time as a function of treatment temperature for the polymer I-5.

For explanation see text.

small extensions ($\sim 0.1\%$) was determined one minute (E_1) and 30 minutes (E_{30}) after the temperature in the working chamber, washed with a stream of argon, had been established. It is seen from Figure 46 that at 80 and 120° the modulus remains practically constant: $\frac{E_{30}}{E_1} \simeq 1$. At 200° the changes connected with imidization are largest, at 250 and 300° the modulus does not markedly change. At 400° there is a large increase, and $\frac{E_{30}}{E_1}$ attains the value of 14 as against the value of 7.5 which corresponds to imidization. It is clear that at 400° the changes in the elasticity parameters due to imidization are no longer significant, and the entire effect is due to secondary transformations.

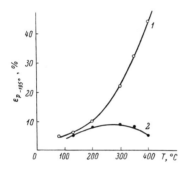

FIGURE 47. Elongation-at-break ε_p at $-195°$C as a function of thermal treatment temperature for polyimides I-5 (1) and VI-5 (2).

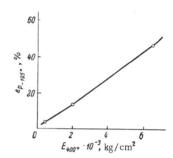

FIGURE 48. Elongation-at-break ε_p at $-195°$C as a function of the static elasticity modulus E measured at 400°C, for polyimide I-5.

We shall now consider the very interesting results of cross-linking in the case of Group C polyimides. As distinct from the cross-linking which usually accompanies degradation processes, the result in this case is an increase rather than decrease of the strength and, most of all, elasticity of the polymer. This is illustrated in Figure 47 which represents the dependence of the elongation-at-break at $-195°$ on the temperature of the preliminary thermal treatment. The treatment temperatures were subsequently raised. The elasticity of polymer I-5 (Group C) markedly increases after heating at 300 and 400°; in the case of polymer VI-5 (Group D), on the contrary, it decreases.

The connection between the cross-linking of Group C polyarimides and the unusual changes in elasticity is shown in Figure 48. It shows the elongation-at-break ε_p at $-195°$ as a function of the static elasticity modulus E at $+400°$. The elasticity modulus could be increased by increasing the duration of heating at 400°. It is seen that the value of the low-temperature elongation-at-break is directly proportional to the elasticity modulus at high temperatures, i.e., to a certain extent to the degree of cross-linking. In this case the density of the samples changed only to an insignificant extent during the thermal treatment; thus, the increase in the elasticity modulus cannot be attributed to crystallization.

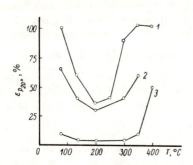

FIGURE 49. Variation of the elongation-at-break ε_p at 20°C after different stages of thermal treatment for films of I-5 polyimide.

Duration of heating at each point — 30 min. 1 — film prepared from fresh solution of polyamido-acid; 2 and 3 — films prepared from solutions preheated at 120°C for 30 and 70 minutes respectively.

It is seen in Figure 49 (curve 1) that the effect of cross-linking may be perceived by testing the elasticity of polymer I-5 films at room temperature as well. During imidization (heating at 200—250°) the elasticity of the films decreases considerably; heating at higher temperatures results in a higher flexibility of the films.

We have already pointed out that when solutions of polyamido-acids are heated at high temperatures, their molecular weight decreases. This is accompanied by partial imidization, and a decrease in the solubility of the polymer. Polyamido-acid films prepared from such solutions have low mechanical strength, and low elongation-at-break (Figure 49, curves 2, 3). The elasticity of the films decreases with increasing temperature and time of preliminary heating of the solution (Figure 49, points on curves 2 and 3 corresponding to 80°). When such films are heated to 200—250° (imidization range) their elasticity decreases — similarly to that of films prepared from fresh solutions (curve 1) — and then, after heating at 300—400° (cross linking range) increases again. This effect is most conspicuous for the most highly degraded sample of polyamido-acid (Figure 49, curve 3).

As has been pointed out above, polyamido-acid films stored in the air for a long time are very weak and brittle, but if subjected to a high temperature treatment, become just as flexible as polyimide films prepared under conventional conditions. These observations and the data represented in Figure 49 show that cross-linking compensates to a large extent for the drop in elasticity produced by the decrease in the molecular weight of the initial polymer due to thermal or hydrolytic chain degradation.

Cross-linking of Group C polyarimides also results in a nonreversible fixation of the oriented state of polyimide /31, 32/. This may be illustrated by comparing the thermomechanical curves of highly oriented capron and polymer PFG (I-9) fibers (Figure 50). Capron fiber contracts strongly near the melting point as a result of removal of the highly elastic deformations produced during stretching; at T_f the fiber increases in length and decomposes. The polyimide fiber shrinks by only 0.5% around 310°, and withstands external loads up to 500° without any further change in dimensions. The same effect is seen in the isometric heating diagrams of fibers made of another representative of Group C — polymer PM (I-5) — as may be seen in Figure 106.

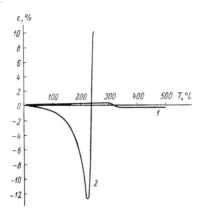

FIGURE 50. Thermomechanical curves of highly oriented fibers.

1 — I-9 polyimide; 2 — capron. Load 20 kg/cm^2, rate of temperature rise 5°/min.

Such effects may in general be observed in any polymer system in which it is possible to fix the oriented macromolecules by means of intermolecular chemical bonds. If the number of cross-links is large, the oriented polymeric system no longer contracts at any temperature and when nonoriented, does not soften, does not undergo forced elastic deformation etc. In cross-linked polyimides the relaxation processes which produce contraction, softening, etc. may not disappear altogether, but may be extended over a wider range of temperatures. This may be seen from Figure 45 which shows that cured PM polymer tends to relax the forced elastic deformations between −200 and +200°, from the thermomechanical curves of this polyimide /19/ on which the parts corresponding to increased deformation shift following thermal treatment, and from the data given below on the measurement of dynamic elasticity modulus at different frequencies.

Figure 51 shows the absolute values of the complex dynamic elasticity modulus of PM polyimide as a function of frequency, determined at 400° and 445° successively. In the latter case the determinations were made 1 and 5 minutes after the temperature had become established (the duration of each determination was about 20 seconds). It is seen that after 5 minutes' heating the elasticity modulus no longer depends on the frequency.

It is interesting to compare the temperature relationships of the dispersion of elasticity modulus* of PM polyimide with those of the typical softening carbon-chain linear polymer — poly(methyl methacrylate) (Figure 52). Whereas in the case of poly(methyl methacrylate) the dispersion has a distinct constant maximum around 130° (curve 1), the dispersion of PM only tends to increase at temperatures above 400°; five minutes' holding at 445° eliminates even this small rise. Thus, cross-linking results, as it were, in spreading the area of dispersion of the

---

* Dispersion may be defined as $D = \dfrac{|E|_{200}}{|E|_{20}}$ where $|E|_{200}$ and $|E|_{20}$ are absolute values of elasticity modulus at 200 and 20 hz respectively.

polyimide over a wider temperature range and in a decrease of the magnitude of the dispersion.

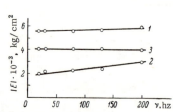

FIGURE 51. Absolute value of complex dynamic elasticity modulus of I-5 polyimide films as a function of frequency.

Determinations made in helium at 400°C, 4 minutes (1) and at 445°C, 1 minute (2) and 5 minutes (3) after temperature had become constant. Films preheated at 300°C.

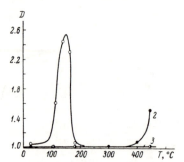

FIGURE 52. Dispersion of elasticity modulus $D = \dfrac{|E|_{200}}{|E|_{20}}$ as a function of temperature.

1 — poly(methyl methacrylate); 2 and 3 — polyimide I-5 heated at 445°C for 1 and 5 minutes respectively. Determinations conducted in helium.

In order to understand the cross-linking mechanism it is interesting to compare the effects of high-temperature changes in polyarimides in different classes, in particular those belonging to Groups C and D. Table 25 shows data on the effect of high-temperature treatment on the mechanical properties of polyimides PM (I-5) and DFO (VI-5) which are typical representatives of Group C and Group D polymers respectively. The tensile strength σ_p and elongation-at-break ε_p of PM and DFO films after heat treatments at 270 and 400° in helium for equal periods of time were determined at +20 and −195°. If we compare the elongations-at-break at −195°, we shall see that the results of heating at 400° are opposite for the two polymers. The elongation of PM is doubled, that of DFO decreases by 25%. The values of elongation-at-break at +20° after heating at 400° do not change much in the case of PM and decrease by one-half in the case of DFO. It also follows from Table 25 that high-temperature treatment results in an increase in the mechanical strength of PM and its decrease in the case of DFO. Thus, DFO displays a deterioration of all the mechanical properties, i.e., a typical consequence of thermal degradation.

TABLE 25. Effect of treatment temperature on the strength σ_p and relative elongation-at-break ε_p for polyarimides I-5 and VI-5

Polymer	Treatment temperature, °C	σ_p, kg/cm²		ε_p, %	
		+ 20°	−195°	+ 20°	−195°
PM (I-5) {	270	1500	2800	65	20
	400	1900	3500	62	45
DFO (VI-5) {	270	2000	2800	100	12
	400	1800	2500	50	8

As distinct from Group C polyimides, Group D polyimides display relaxation effects in a narrow temperature range only, which does not significantly vary with the conditions of the thermal treatment. These polymers have a distinct range of strong dispersion of modulus of elasticity, and the temperature dependence of the dispersion may be seen from Figure 53 to be fully similar to that noted in linear carbon-chain polymers (cf. with Figure 52).

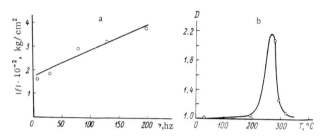

FIGURE 53. Relaxation characteristics of polyimide VI-5.

a — absolute dynamic elasticity modulus as a function of frequency at 274°C; b — dispersion of elasticity modulus $D = \frac{|E|_{120}}{|E|_{10}}$ as a function of temperature. Experiments performed in argon.

The high elasticity of some of Group C polyimides at low temperatures and the absence of distinct dispersion range of elasticity modulus in these polyimides are probably interconnected. It is known that all relaxational properties of polymers are most clearly displayed in the dispersion range. At temperatures below the dispersion range of elasticity modulus the polymers usually behave as solid, brittle bodies. In Group C polyimides the dispersion is spread over a wide temperature range, in which relaxational effects are free to develop. This accounts for effects such as the large deformations consequent on stretching PM polymer between −200 and +400°, since these are due to the possibility of continuous stress relaxation during the deformation. Thus, the structure acquired by Group C polyimides as a result of high-temperature treatment is in a sense equivalent to a mechanical model consisting of elements with a large spectrum of times of relaxation. In Group D polyimides, on the other hand, this spectrum is narrow.

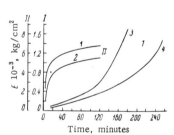

FIGURE 54. Elasticity modulus as a function of time during high temperature treatment of Group C and D polyimides.

Films preliminarily heated at 300°C. Experiments performed in argon. Polyimide I-5; 1 — 386°C; 2 — 398°C; polyimide VI-5: 3 — 424°C; 4 — 414°C.

These differences affect not only the results of cross-linking processes in polyimides of Groups C and D, but also the kinetics of these processes. Figure 54. shows the time dependence of elasticity moduli of polymer films PM (I-5) and DFO (VI-5) held in argon at high temperatures. The films had been pretreated in argon at 300° for 1 hour. It is seen that in the former case the variation of elasticity

89

modulus E with time t may be expressed by the formula $E \simeq A\,(1-e^{-\alpha t})$, whereas in the latter it proceeds according to formula $E \simeq Be^{\beta t}$, as can be confirmed by the semilog graphs in Figure 55. The rate of variation of the elasticity modulus (especially in the initial stages) is much higher for polyimide PM than for polyimide DFO. It should be noted that the elasticity modulus values of polyimide PM are higher at lower temperatures (Figure 54, curve 1), whereas in the case of polyimide DFO (curve 3) higher values of elasticity modulus correspond to higher temperatures. The difference consists in the fact that DFO is in a highly elastic state at high temperatures (i. e., the elasticity modulus increases with temperature), whereas the elasticity of PM is due to a considerable extent to intermolecular forces, which become weaker at higher temperatures.

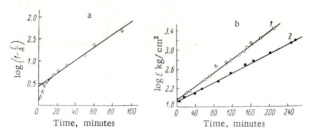

FIGURE 55. Dependence of $\log \left(1 - \dfrac{E}{A}\right)$ on time for polyimide I-5 (a) at 386°C and of $\log E$ on time for polyimide VI-5 (b) at 424°C (1) and 414°C (2). Experiments carried out in argon.

Graphs such as those shown in Figure 54 and 55 can be utilized to calculate the variation rates $\dfrac{dE}{dt}$ of elasticity modulus with time during different stages of thermal treatment at different temperatures. If the treatment time t is kept constant, the dependence of $\log \left(\dfrac{dE}{dt}\right)$ on the reciprocal absolute temperature $\dfrac{1}{T}$ will be represented by a straight line (Figure 56), in accordance with Arrhenius' equation

$$k = Ae^{-\frac{U}{RT}},$$

which can be utilized to calculate the activation energy U of processes responsible for these temporal changes in the elasticity modulus. In both cases the activation energy proved to be variable (Figure 57). The activation energy of polyimide PM (a Group C polymer) is low (6—7 kcal/mole) in the initial reaction stages, when the variations in the elasticity modulus values are strongest; high activation energies (30—50 kcal/mole) are noted only in the advanced stages of thermal treatment. The activation energies of polyimide DFO (a Group D polymer) are much higher, increase more rapidly with time and attain 70—80 kcal/mole in the advanced stages of the treatment. It has already been pointed out (Chapter II, p.39) that such values are typical of the activation energies of thermal degradation of polyimides.

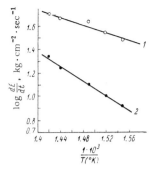

FIGURE 56. Logarithm of the rate
of increase of elasticity modulus
as a function of reciprocal absolute
temperature at different stages of
cross-linking of polyimide I-5.

$1 - t = 15-20$ min; $2 - t =$
$50-70$ min. Experiments conducted in argon.

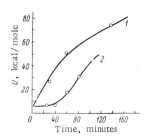

FIGURE 57. Variation of activation energy U of cross-linking
of polyimides VI-5 (1) and I-5
(2) as a function of the duration
of thermal treatment.

It may thus be concluded that at least two different cross-linking
processes take place. One of them has low activation energies, is typical
of Group C polyimides — polypyromellitimides — and results in a marked
improvement of mechanical properties. The other process has large
activation energies and results in a deterioration of the mechanical
properties of the polymer. It is probably connected with irreversible
degradation of chain macromolecules resulting from thermal decomposition. It is possible that the reason why this degradative cross-linking
mechanism is mainly operative in Group D polymers is that at high
(about 400°) temperatures these polymers are mostly in a highly elastic
state, but such type of cross-linking is also noted in Group C polyimides
(polypyromellitimides), especially if these have been exposed to high
temperatures for long periods of time. One example is the deterioration
of the mechanical properties of "H-film" during prolonged aging
(Chapter IV, p.121) /69/ from which it was concluded that the activation
energy of the process responsible for the decrease in mechanical strength,
elasticity, etc. with time is 55 kcal/mole, i.e., is close to the activation
energy of thermal degradation and in close agreement with cross-linking
energies of PM at advanced stages of thermal treatment.

The mechanism of the former, specific type of cross-linking was
discussed /31/ by the authors of this book, according to whom it consisted
in the formation of long intermolecular bonds as a result of cleavage
of imide rings and recombination. If this point of view is accepted, the
differences in the processes involving Group C and Group D polyimides
may be attributed to a possible variation of the reactivity of the rings with
the type of links between them.

Temporal variations of elasticity modulus at high temperatures —
usually an increase — are noted for most polymers of different structures.
Occasionally, e. g., in crystalline polymers, this effect may be produced
solely by physical factors (crystallization). Potentially crystalline polyimides in fact display such effects (p.110). The polymers which have

been discussed above do not crystallize, so that the changes in their elasticity moduli can be explained only in terms of chemical processes. This may be demonstrated directly by studying the changes of elasticity modulus in different gaseous media. Figure 58 shows the results of such measurements for III-9 and I-5 polyimides in helium and in air. It is seen that in the presence of oxygen the rate of cross-linking immediately increases by a factor of several dozen. Clearly, if the variations in the modulus had been due to crystallization only, such an effect could not have been observed.

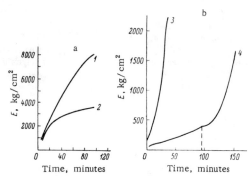

FIGURE 58. Temporal variation of elasticity moduli of polyimides I-5 (a) and III-9 (b) at 400°C.

1,3 — in the air; 2 — in argon; 4 — in helium (after 97 minutes the helium was replaced by air)

It may be said in conclusion that high-quality polyimide materials can only be obtained if the secondary chemical processes which take place in them at high temperatures are allowed for, since such processes may in some cases improve the technological parameters of the final product and in others, on the contrary, produce deterioration.

Group A and B polyimides have not been discussed in this section. It is seen from Table 24 that the elasticity moduli of these polyimides show no marked changes at high temperatures. The values of elasticity modulus at 400° and at higher temperatures are very high. Very probably, cross-linking processes are here masked by the high rigidity of the chains and a strong intermolecular interaction between large groupings with a flat configuration. Accordingly, the properties of these systems are mainly determined by the initial structure of the monomer link and the degree of completion of the imidization and not by secondary chemical reactions.

Effect of chemical structure on the physical properties of polyimides belonging to different groups

We shall now consider the physical properties of the individual representatives of the four classes of aromatic polyimides as defined above.

This will make it possible to correlate these properties with the special features of chemical structure within each group. It will be seen, moreover, that some of the polyimides display properties which are intermediate rather than typical of a particular group. This indicates that the classification given above is not perfect and is also indicative of the general properties of this polymer class as a whole.

Group A polyimides (Table 24) do not contain links which reduce the energy barrier of rotation around single bonds contiguous to the ring groupings. The flat shape of the monomeric units is responsible for the considerable rigidity of the macromolecules, strong intermolecular van der Waals forces and a high packing density. Figure 59 represents the elasticity modulus of a number of Group A polyimides as a function of the temperature. The values of elasticity modulus are typically high, both at room temperature ($\sim 10^5$ kg/cm^2) and up to 400° ($10^4 - 10^5$ kg/cm^2). It is also seen that as the number of phenyl nuclei in the links increases, the variation of the modulus with the temperature becomes stronger. The mechanical strength and especially the elasticity increase at the same time (Table 24). These effects may be attributed to a certain decrease in the rigidity of the chain. The high values of elasticity modulus are due to strong intramolecular and intermolecular forces. Against this background, the changes in the elasticity modulus due to cross-linking are insignificant. In fact, the elasticity modulus of Group A polymers increases by only 10—50% during 15 minutes at 400°, whereas in Group C polymers the corresponding increase is considerably greater (Table 24). Even though softening effects are totally absent, Group A polyimides are capable of forced elastic deformation at high temperatures, as may be seen from the high-temperature tensile diagrams, e.g., for polymer I-2 film at 400° (Figure 60).

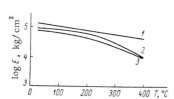

FIGURE 59. Logarithm of elasticity modulus as a function of temperature for a number of Group A polyimides.

1 — I-2; 2 — III-1; 3 — III-2.

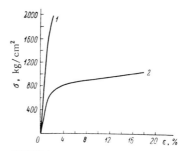

FIGURE 60. Tensile diagrams of polyimide I-2 at 20°C (1) and at 400°C (2).

Polyarimides in this group display the largest thermal stability. Thermooxidative processes affect the mechanical properties to a lesser extent than in the case of polyimides belonging to other groups. Thus, polymer III-2 (Group A) has in the initial state an elongation-at-break 10 times smaller than polymer I-5 (Group C), for an approximately equal mechanical strength. After aging in the air at 400° for 3 hours polymers III-2 and I-5 acquire comparable elongations-at-break, while polymer III-2 becomes twice as strong.

Group B polyimides contain heteroatomic groupings in the dianhydride radical only. However, the "hinge" links, even if their number
is fairly large, do not ensure marked ductility, owing to the fact that the
contiguous flat-shaped groupings are very
bulky. Accordingly, the properties of
Group B polyimides resemble those of
Group A. These polymers also display
low elasticity which have high elasticity modulus at 20° and high density (Tables 21 and 24).
The temperature dependence of the elasticity modulus, however, is more marked
than for Group A polymers (Figure 61,
cf. Figure 59), apparently due to the
fact that some rotation around the "hinges"
is possible at high temperatures and that
the intermolecular forces are correspond-
ingly weaker. Nevertheless, the elasticity
moduli remain high, even at 400° (about
10^4 kg/cm^2). Group B polyimides do not
soften. Cross-linking produces somewhat

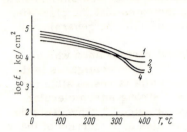

FIGURE 61. Logarithm of elasticity
modulus as a function of temperature
for a number of Group B polyimides.

1 — VI-1; 2 — VII-2; 3 — IV-2;
4 — VI-2.

greater changes in elasticity modulus (Table 24) and the mechanical
properties deteriorate somewhat more rapidly than those of Group A
polymer as a result of aging in the air.

It must be noted that these properties are displayed by ring-chain
systems,· the R' units of which contain benzene rings joined exclusively
in the para-position. If the links between the rings are in the meta-
position, the energy barrier of rotation becomes lower and the resulting
effect is similar to that produced by the introduction of a heteroatom.
For example, polymer VI-1 (R' — p-phenylene) has an elasticity modulus
$E = 2 \cdot 10^4$ kg/cm^2 at 300°, whereas polymer VI-3 (R' — m-phenylene) has
$E = 50$ kg/cm^2. It is more correct, accordingly, to assign the latter
compound to Group D polyimides with "hinges" in both components of the
monomeric unit, as has in fact been done in Table 24.

'Group C polyimides have a short, rigid nucleus R, e.g., pyro-

mellitic acid residue ⟨structure⟩ and "hinge" hetero-atom groups between the

phenyl nuclei in the diamine component, e. g. ⟨structure⟩–0–⟨structure⟩– or

⟨structure⟩–0–⟨structure⟩–0–⟨structure⟩– . Such a structure greatly facilitates the
rotation of all elements in the chain, including aromatic nuclei with conjugated imide rings, relative to one another. Group C polymers, especially
derivatives of pyromellitic dianhydride, thus acquire an ensemble of
technologically very valuable properties. The higher elasticity of these
polymers is combined with considerable hardness and mechanical strength
at both high and low temperatures. Intermolecular forces are much weaker
than in Group A and Group B polyimides. The elasticity modulus of
Group C polymers at 400° is $1 \cdot 10^2 - 5 \cdot 10^2$ kg/cm^2 as compared with
some 10^4 kg/cm^2 for Group A polyimides. However, molecular network
formation produces a marked rise in the elasticity modulus at 400° to
$2 \cdot 10^3 - 6 \cdot 10^3$ kg/cm^2 (cf. Figures 18, 54, and others).

Figure 62 shows the elasticity modulus as a function of the temperature for a number of Group C polyimides based on dianhydrides of pyromellitic (I-5, I-9, I-7) and 2,3,5,6-pyridinctetracarboxylic (II-10, II-5, II-9, II-7) acids and different aromatic diamines. In both kinds of polymers the imide rings are rigidly connected by means of radical R. It is seen that if radical R' has the same structure, the temperature dependence of the modulus is similar in both cases. However, the marked change in the modulus (softening range) corresponds to higher temperatures in the case of polypyromellitimides than in the case of polypyridine-imides. In both cases an increased number of "hinges" in R' shifts the drop range of elasticity modulus towards lower temperatures (cf. I-5 and I-9, II-5 and II-9). Polymers with a sulfide group —S— in the chain show a more rapid change of elasticity modulus with the temperature than do polymers with an ether group —O— (cf. I-5 and I-7, II-5 and II-7). If the phenyl ring in R' is located in the meta-position and there are two "hinge" ether groups in this radical, the variation of elasticity modulus in the softening range remains largely unaffected (cf. II-9 and II-10). Effects of high-temperature cross-linking are considerable, especially if the polymer displays a marked tendency to softening. This is easily seen in Figure 62 (range of increase in the modulus with temperature, especially for I-7 and II-7) and can also be deduced from literature data (Figure 63).

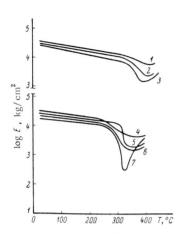

FIGURE 62. Logarithm of elasticity modulus as a function of the temperature for Group C polyimides.

1 — I-5; 2 — I-9; 3 — I-7; 4 — II-10; 5 — II-5; 6 — II-9; 7 — II-7.

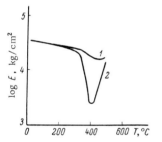

FIGURE 63. Logarithm of elasticity modulus as a function of temperature for polypyromellitimides with

$$R' = -\langle \bigcirc \rangle - O - \langle \bigcirc \rangle - \quad (1)$$

and

$$R' = -\langle \bigcirc \rangle - NH - CO - \langle \bigcirc \rangle - \quad (2).$$

After /58/.

After complete thermal treatment which imparts to them the optimum properties, polypyromellitimides lose their capacity to soften and to undergo other thermal transitions /35, 36/, the bends in thermomechanical curves are shifted to higher temperatures /19/, and the capacity to undergo forced elastic deformation is extended to a wider temperature range.

Group C polyimides based on 3,3',4,4'-diphenyltetracarboxylic acid dianhydride, in which the imide rings are conjugated with two phenyl rings interconnected by a single bond, soften much more readily than do polypyromellitimides. It is seen from Figure 64, which gives the elasticity moduli of a number of such polyimides as a function of the temperature, that the softening capacity increases with the increasing number of "hinge" groups in the monomer unit (cf. III-5 and III-9) and is higher when the bonds are in the meta than in the para position (cf. III-9 and III-10).

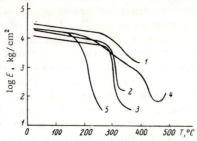

FIGURE 64. Logarithm of elasticity modulus as a function of the temperature for a number of polyimides based on the dianhydride of 3,3',4,4'-diphenyltetracarboxylic acid.

1 — III-5; 2 — III-8; 3 — III-3; 4 — III-9; 5 — III-10.

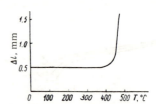

FIGURE 65. Thermomechanical curve of plastic material made of polyimide III-9.

Load $100 \, kg/cm^2$, heatup rate 5 degrees/minute.

It would appear that the softening capacity is due to the increasing mobility of imide ring groups around the diphenyl bond at high temperatures. This mobility seems to suffice not merely to produce softening but also to bring about crystallization; this is noted, for example, when polymer III-10 is heated above 250°, which obviously involves a major rearrangement in the mode of packing of the chains. In a number of cases the softening capacity of polyarimides based on 3,3',3,4'-diphenyltetracarboxylic acid dianhydride is not lost even after heating at high temperatures. This is seen, for example, on the thermomechanical curve obtained for a sample of plastic material obtained by subjecting polymer III-9 at 460° to a pressure of $\sim 2000 \, kg/cm^2$ (Figure 65). The shape of the curve is typical of linear polymers. This means that the cross-linking effects are here weaker than in the main representatives of Group C — polypyromellitimides. Polyimide derivatives of 3,3',4,4'-diphenyltetracarboxylic acid with "hinge" groups in the diamine component display properties similar to those of Group D polyarimides.

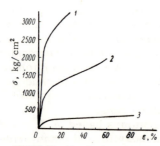

FIGURE 66. Tensile diagrams of I-5 polyimide films at different temperatures.

1 — −195°; 2 — 20°; 3 — 400°.

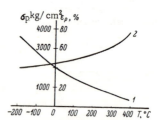

FIGURE 67. Tensile strength (1) and elongation-at-break (2) as a function of the temperature for polyimide I-5 film.

It has been pointed out above that many Group C polyimides have superior stress-strain properties which are very valuable from the technological point of view. A typical example of such a polymer is polypyromellitimide PM (I-5), the film tensile diagrams of which are shown in Figure 66, while Figure 67 shows its tensile strength σ_p and elongation-at-break ε_p as a function of temperature. It is seen that this polymer retains high strength and ductility in a very wide temperature range — between —200 and +400°.

A conspicuous feature is the retention of ductility at cryogenic temperatures (up to the temperature of liquid helium /38/). The ductility is retained not only during slow deformations realized in conventional laboratory testing procedures (stretching rate of about 1%/sec), but also at fast stretching rates (10^5 —10^6 %/sec). For the sake of comparison it may be noted that Teflon and Terylene films display a brittle rupture even at 20° at such high stretching rates, whereas the rupture of polymer PM is elastic both at +20° and at —195° It may also be pointed out that the purely elastic deformation range on PM tensile diagrams attains elongations up to $\varepsilon = 5 - 6\%$, whereas in the range of forced elastic deformation (flatter parts of diagrams in Figure 66) the elasticity modulus increases considerably.

Table 26 lists data on the strength σ_p and elongation-at-break ε_p of Group C polyimides.

In this connection we may mention certain typical relationships. If the —O— ether group is replaced by the sulfide group —S—, the strength and elasticity decrease to varying extents (I-5 and I-7, II-5, II-7). An increase in the number of ether bonds results in an increase in elasticity (I-5 and I-9, II-5 and II-9, III-5 and III-9). Ether or ester bonds in conjunction with a phenyl nucleus in the meta-position (I-10 and I-13), on the contrary, reduce the strength and elasticity of the product. The limited mobility around the diphenyl bond at low temperatures reduces the elasticity of polyimides based on the dianhydride of 3,3',4,4'-diphenyl-tetracarboxylic acid by a factor of 5 —10 as compared to polypyromellit-imides. At high temperatures, on the other hand, the mobility is responsible for the greater deformability.

The effect of thermal and thermooxidative degradation on the mechanical properties of Group C polyarimides depends on the nature and number of groups in the monomeric unit. Polyimides based on 2,3,5,6-pyridinetetra-carboxylic acid are much less thermostable than polypyromellitimides. In polypyromellitimides the rate of decrease of elasticity as a result of heating in the air (Table 27) and in an inert medium increases with increasing number of hetero-atoms in the diamine. The rate of decrease in tensile strength varies to a smaller extent. Data on the variation of strength and elasticity of oriented and nonoriented polymer PM films in the air and nonoriented films in helium are also shown in Figure 68 and Table 22 respectively.

Heacock and Berr /69/ conducted a detailed study of the thermostability of physical properties of "H-film" — a commercial product based on polyimide PM (cf. Chapter IV).

Group D polyimides have "hinge" groups in both the dianhydride and the diamine components. The groupings between the "hinges" are small in size. Owing to these structural features Group D polyimides, similarly to the linear carbon-chain polymers, become highly elastic in a narrow temperature range and have a melting point. The shape of the curves representing the elasticity modulus as a function of temperature

TABLE 26. Strength properties of Group C polyimides

Formula of polymer	Symbol of designation of polymer	−195°		+20°		+200°		+400°	
		σ_p, kg/cm^2	ε_p, %	σ_p, kg/cm^2	ε_p, %	σ_p, kg/cm^2	ε_p, %	σ_p, kg/cm^2	ε_p, %
[structural formula]	I-5	3600	45	2000	100	1000	47	500	120
[structural formula]	I-7	2000	8	1300	30	700	24	—	—
[structural formula]	I-9	3500	30	2000	130	—	—	600	150
[structural formula]	I-10	—	—	600	6	—	—	—	—
[structural formula]	I-13	—	—	800	4	—	—	150	1
[structural formula]	II-5	—	—	1500	10	—	—	—	—
[structural formula]	II-7	—	—	1000	6	—	—	—	—
[structural formula]	II-9	—	—	1650	50	—	—	—	—
[structural formula]	II-10	—	—	1500	7	—	—	—	—
[structural formula]	III-5	—	—	1400	15	—	—	—	—
[structural formula]	III-7	—	—	1400	15	—	—	—	—
[structural formula]	III-9	—	—	1800	20	—	—	—	—
[structural formula]	III-10	—	—	1800	15	—	—	—	—

TABLE 27. Effect of thermooxidative degradation on the strength parameters of polypyromellitimides PM
(I-5) and PFG (I-9)
Isotropic films 30 μ thick; tested at 20°C

Conditions of degradation		I-5		I-9	
temperature, °C	time, hours	σ_p, kg/cm^2	ε_p, %	σ_p, kg/cm^2	ε_p, %
Initial samples		1600	81	1600	100
350	2	1630	58	1610	44
350	5	1580	33	1600	20
400	1	1415	30	1500	12
400	2	620	4.4	780	4.4
400	3	620	2.6	500	2

(Figures 69 and 70, see also Figure 44) and the course of temperature
dispersion of elasticity modulus (cf. Figures 52, 53) confirm this effect.

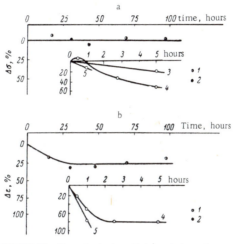

FIGURE 68. Decrease in strength (a) and elongation-
at-break (b) at 20°C, in % of initial values for films
of polyimide I-5 as a result of thermal aging.

1— oriented, 300°C, in the air; 2— oriented, 400°C,
in argon; 3 — oriented, 400°C, in the air; 4 — nonoriented, 400°C, in the air; 5 — nonoriented, 500°C,
in the air.

The range of transition to the softened state will depend on the number
and structure of the groupings which make internal rotation in macromolecules possible. For example, if the sulfide group —S— in the diamine
is replaced by the disulfide group —S—S—, the softening temperature
decreases (VI-7, and VI-8 in Figure 69). A similar result is produced
if the phenyl ring between the ether bonds is moved from para to meta
position (VI-9, Figure 81 and VI-10, Figure 69). If ether groups are
replaced by sulfide groups, the softening temperatures will remain
practically unchanged (VI-5, Figure 44 and VI-7, Figure 69). A parallel
change in Group C polyimides resulted in a considerable enhancement in
softening tendency.

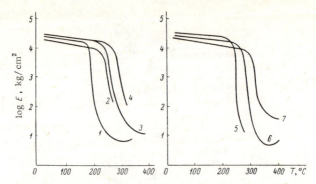

FIGURE 69. Logarithm of elasticity modulus as a function of temperature for Group D polyimides:

1— VI-10; 2— VI-8; 3— VI-7; 4— IV-12; 5— VII-3; 6— VII-4; 7— VI-4.

The groups in R and R' which connect the phenyl nuclei and which are contiguous to the imide ring may be arranged in the following sequence of increasing softening temperatures:

$$-O-\text{⟨⟩}-O-, \quad -O-\text{⟨⟩}-O-, \quad -S-S-, \quad -S-,$$

$$-O-, \quad -CO-, \quad -SO_2-.$$

The highest softening temperature (about 500°) is displayed by polyimide V-4, which contains sulfone groups $-SO_2-$ in R and R'. However, polyimides with $-SO_2-$ groups in the chain usually have poor mechanical properties owing to their low molecular weight. The ready softening and melting displayed by polyarimides of Group D make it possible to work them into products of the thermoplastic type. Such products often have superior physical and mechanical properties (Chapter IV) and high softening points (Figure 71).

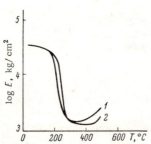

FIGURE 70. Elasticity modulus as a function of temperature for polyimides based on the dianhydride of 3,3',4,4'-benzophenone-tetracarboxylic acid /58/.

1— IV-3; 2— IV-5.

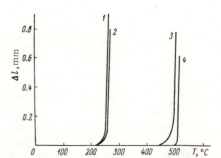

FIGURE 71. Thermomechanical curves of plastic products made of Group D polyimides.

Load 100 kg/cm²; rate of heating 5 deg/min.
I — VII-3; 2 — VI-5; 3 — VI-3; 4 — V-4.

100

Cross-linking of Group D polyimides takes place in the highly elastic state. The related changes in the elasticity modulus are, as a rule, small (Table 24). The softening range becomes somewhat wider after heating at high temperatures, but does not shift to a significant extent. Polyimides in which m-phenylene constitutes the R'-part form an exception. Here cross-linking proceeds at an intense rate, the elasticity modulus increases to 10^4 kg/cm^2 at 400° and the softening temperature undergoes a marked shift. Crystallization may take place in the softened state. Data on the temperature dependence of elasticity modulus of DFM polymer film (VI-3) preheated to 250° are shown in Figure 72, as an example. On heating there appears a softening range around 260°, a crystallization range (300—400°) in which the elasticity modulus increases, and a melting range around 490—500°. At high temperatures the polymer becomes rapidly cross-linked. The result is that on being cooled (reverse path along the curve) the polymer remains amorphous, no longer softens at 260° and has higher values of elasticity modulus than during heatup at all temperatures below 400°. If the polymer is now reheated, the only effect which is reproduced is the variation of the modulus with the temperature represented by the reverse (cooling) curve in Figure 72. After heating at high temperatures, the softening range shifts from 260 to 490°, as may be seen, for example, from Figure 71 which represents the thermomechanical curve of a DFM plastic product compressed at 500°. We shall see in what follows, when dealing with crystallization of a number of other polyarimides of Group D with a large number of "hinges" in the chain, that cross-linking, just as in the case of DFM, takes place above the melting point, prevents formation of crystalline state, but does not affect the softening range and produces only a slight increase in the elasticity modulus.

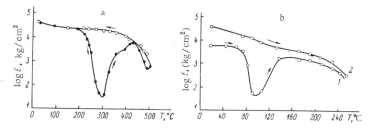

FIGURE 72. Logarithm of elasticity modulus as a function of the temperature for film of polyimide VI-3 preheated to 250°C (a), and film of poly(ethylene terephthalate) (b).

— heatup; o — cooling. 1 — quenched amorphous film; 2 — crystallized film.

Many Group D polyimides are very strong and elastic. Elongation-at-break at 20° may be as high as 100%, similar to that of Group C polyimides. Large forced elastic deformations develop owing to the ease with which rotation around monomeric link elements may be realized because of the separation of the phenyl nuclei and polyimide rings by "hinge" groups on both sides. Figure 73 gives tensile diagrams of DFO (VI-5) polymers at different temperatures. At 20° a fairly large range of forced elastic

TABLE 28. Physical and mechanical properties of Group D polyimides

Formula of polymer	Designation of polymer	σ_p, kg/cm^2 (20° C)	ε_p, % (20° C)	T_s, °C	T_f, °C
[structure]	IV-5	1600	15	290	—
[structure]	IV-9	1600	25	270	—
[structure]	V-5 *	—	1	340	—
[structure]	V-4 *	—	1	~500	—
[structure]	VI-3 **	1500	30	~260	~490
[structure]	VI-3 †	1250	5	460	—
[structure]	VI-4 *	—	1	310	—
[structure]	VI-5	2000	100	270	—
[structure]	VI-6	1200	20	250	—
[structure]	VI-7	1200	30	265	—
[structure]	VI-8	1000	9	230	—
[structure]	VI-9 ***	1400	50	250	~420
[structure]	VI-10	1300	25	200	—
[structure]	VI-12	1300	30	290	—
[structure]	VII-3	1400	40	260	—
[structure]	VII-4	1100	15	290	—
[structure]	VII-5	1100	40	220	—
[structure]	VII-9 **	1300	90	200	~390

* Low-molecular weight polymer.
** Data obtained for amorphous film.
† Data obtained for plastic product, heated above T_f

deformations may be noted. On passing the limiting forced elasticity a distinct peak is noted. It may be noted that this peak does not appear on the tensile diagrams of Group C polyimides (e. g., PM, Figure 66). At −150°, DFO still displays large values of ε_p, but at −195° the elongations-at-break are several times smaller than those given by PM, even though at +20° DFO may have higher values of tensile strength and elongation-at-break. Figure 74 shows temperature relationships of strength and elongation-at-break for films of this polymer.

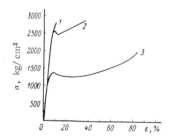

FIGURE 73. Tensile diagrams of film of polyimide VI-5 at different temperatures.

1 — −195°; 2 — −150°; 3 — 20°.

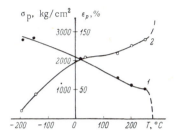

FIGURE 74. Temperature dependence of tensile strength (1) and elongation-at-break (2) for a film of polyimide VI-5.

It may be noted that, as distinct from Group C polyimides, cross-linking of Group D polyimides results in decreased elasticity (Figure 47).

Table 28 shows thermal and mechanical deformation data for a number of Group D polyimides.

It is noteworthy that, similarly to Group C polyimides, polymers with a sulfide bond —S— in the diamine are less strong and elastic, but soften somewhat more readily than those with an ether group —O— (VI-5 and VI-7). The softening temperature shows a marked decrease if the phenyl ring is shifted from para (VI-9) to meta position (VI-10). Crystalline films are much less elastic than the amorphous. The extreme brittleness of certain polyimides with an —SO$_2$— group in the chain (VI-4, V-5, V-4) is due in the first place to the low molecular weight of the polyamide-acids. A higher-molecular polymer of this type (VII-4) yields polyimide films with satisfactory properties.

Group D polyimides without aliphatic links are highly resistant to thermooxidative degradation. It is seen from Table 29 that their tensile strength σ_p and elongation-at-break ε_p practically do not decrease, when the polyimides are held at 380° for one hour in the air. This property is very important in technological processing.

After being heated in helium at 400° for 3 hours, the films retain fairly high values of σ_p and ε_p. It should also be noted that individual polymers within the group display very different rates of thermal degradation with consequent alteration in their physical and mechanical properties. A large number of ether bonds or extension of the monomeric unit produce a marked increase in the rate of degradation. A similar effect is observed on introducing one or two sulfone groups (Figure 35).

TABLE 29. Effect of thermal degradation on the mechanical properties of Group D polyimides at 20°C. Film thickness 30—40 μ, films preheated at 300°C in vacuo for one hour

Conditions of degradation		IV-5		VI-5		VII-3		VII-4	
temperature, °C, and medium	time, hours	σ_p, kg/cm^2	ε_p, %	σ_p, kg/cm^2	ε_p, %	σ_p, kg/cm^2	ε_p, %	σ_p, kg/cm^2	ε_p, %
Initial samples		1600	16	1400	67	1400	57	1150	16
380°C, in the air	1	—	—	1400	62	1500	50	1200	16
380°C, in the air	3	—	—	1360	29	1500	17	1100	10
400°C, in helium	3	1400	7	1400	25	—	—	—	—

Certain properties of polyimides based on nonaromatic 1,2,3,4-butane-tetracarboxylic and 2,3,4,5-cyclopentanetetracarboxylic acid dianhydrides (VIII and IX in Table 19) with corresponding aromatic diamines justify their assignment to Groups B and D. In 1,2,3,4-butanetetracarboxylic acid rotation around $-C-C-$ single bond is possible. The grouping formed by the imide rings and the cyclopentane ring is not flat.

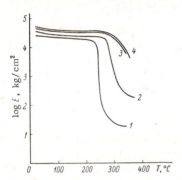

FIGURE 75. Logarithm of elasticity modulus as a function of temperature for polyimide films based on 1,2,3,4-butanetetracarboxylic acid dianhydrides.

I — IX-9; 2 — IX-5; 3 — IX-2; 4 — IX-1.

Figure 75 shows the elasticity modulus as a function of the temperature for several polymers based on 1,2,3,4-butanetetracarboxylic acid. A comparison of the figure with similar relationships for polymers based on 3,3',4,4'-tetracarboxydiphenyl oxide dianhydride shows that polyimide obtained from 1,2,3,4-butanetetracarboxylic acid and rigid diamines such as benzidine (IX-2) softens somewhat more readily than the corresponding Group B polyimide (VI-2, cf. Figure 61). This is probably due to the fact that mobility around the "hinge" in radical R is more strongly inhibited in the case of aromatic Group B polyimides owing to the rigid structure of radical R' and the larger size of radical R than in the corresponding derivatives of 1,2,3,4-butanetetracarboxylic acid. When in combination with flexible diamines, the softening temperatures of polyimides based on 1,2,3,4-butane-tetracarboxylic acid are always somewhat higher than those of the corresponding aromatic Group D polyimides (cf. IX-5, Figure 75 with VI-5, Figure 44, and IX-9, Figure 75 with VI-9, Figure 78).

Softening temperatures of corresponding polyimides based on 2,3,4,5-cyclopentanetetracarboxylic acid dianhydride are higher than those of polyimide derivatives of 1,2,3,4-butanetetracarboxylic acid. For example, when in combination with 4,4'-diaminodiphenyl ether, these dianhydrides give polymers with respective softening points of 290 and 325° (Table 30).

Figure 76 shows thermomechanical curves of plastic products made of polyimides on the base of 1,2,3,4-butane- and 2,3,4,5-cyclopentanetetra-carboxylic acid dianhydrides and flexible diamines. It is seen that softening takes place in a narrow temperature range, as is typical of Group D polyimides. As in the case of aromatic dianhydrides, an increase

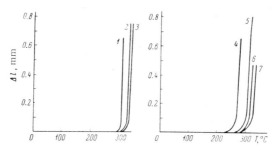

FIGURE 76. Thermomechanical curves of plastic products made of polyimides based on 2,3,4,5-cyclopentanetetracarboxylic acid and 1,2,3,4-butanetetracarboxylic acid.

Load 100 kg/cm^2; rate of heating 5 degrees/minute.
1 — VIII-9; 2 — VIII-3; 3 — VIII-5; 4 — IX-9; 5 — IX-3;
6 — IX-7; 7 — IX-5.

in the number of ether groups and the replacement of the ether bond in the diamine by a sulfide bond result in a decrease of the softening point.

TABLE 30. Properties of polyimides based on 1,2,3,4-butane- and 2,3,4,5-cyclopentanetetracarboxylic acid dianhydrides

Polymer	At 20°		At 250°		Softening point, °C
	σ_p, kg/cm^2	ε_p, %	σ_p, kg/cm^2	ε_p, %	
VIII-5*	—	—	—	—	325
VIII-9*	1000	50	—	—	315
IX-2	1200	10	650 (350**)	3 (50**)	—
IX-5	1200	15	620	10	290
IX-9	1100	40	450	180	270

* Data obtained for pressed samples.
** At 350°C.

Polyimides obtained from these anhydrides are amorphous and their density is lower than that of the corresponding aromatic products. Some deformation and mechanical parameters of this group of polymers are shown in Table 30. The mechanical properties of polyimides based on 1,2,3,4-butane- and 2,3,4,5-cyclopentanetetracarboxylic acid dianhydrides deteriorate as a result of thermooxidative degradation at a much faster rate than do those of polyimide derivatives of aromatic tetracarboxylic acids. For example, when aged in the air at 300° for 6 hours, the tensile strength and elongation-at-break of IX-2 polyimide film decrease to about one-half. The film loses 5% of its own weight when heated in the air at 300° for 2 hours and 50% when heated at that temperature in the air for 24 hours. Aromatic polyimides lose practically no weight and do not alter their physical and mechanical parameters as a result of such a treatment. It should be pointed out, however, that these polymers, while less thermostable than the aromatic polyimides, are still much superior in this respect to carbon-chain polymers.

Copolyimides

Copolycondensation is a well-known procedure for regulating the supermolecular structure and properties of polyimides. Copolymers of polyimides (copolyimides) may be prepared by condensation of more than one dianhydride with one diamine or more than one diamine with one dianhydride. The preparation of a number of copolyimides has been described in patent literature (see, for example, No. 41 and No. 88 in the list of patents, Appendix I). The following may be noted as a result of systematic comparison of homo- and copolyimides.

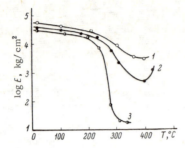

FIGURE 77. Logarithm of elasticity modulus as a function of the temperature for polyimides VI-2 (1), VI-5 (3) and of the copolyimide of the dianhydride VI with diamines 2 and 5, taken in the ratio 1:1:1 (2).

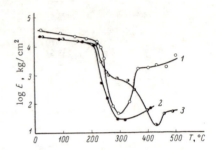

FIGURE 78. Logarithm of elasticity modulus as a function of the temperature for polyimides VI-3 (1), VI-9 (3) and copolyimide of the dianhydride VI with diamines 3 and 9, taken in the ratio 1:1:1 (2).

Copolyimides of dianhydrides and diamines yielding polyarimides of Groups A, B and C are not very different from the corresponding homopolymers.

If the dianhydride forms a Group D polyimide (which has a softening point) with one of the diamines and a Group B (nonsoftening) polyimide with the other, the copolymer of the anhydride with both diamines displays properties which are intermediate between those of the respective homopolyimides. This can be seen, for example, in Figure 77, which shows the elasticity modulus as a function of the temperature for the copolymer of the dianhydride of 3,3',4,4'-tetracarboxydiphenyl oxide (R = VI) with benzidine (R' = 2) and 4,4'-diaminodiphenyl ether (R' = 5) taken in equimolecular proportions (1:1:1) and for the respective homopolymers VI-2 and VI-5.

If a given dianhydride gives crystallizing, softening Group D polyimides with each of the two diamines separately, the softening temperature of the copolymer will be somewhat lower than the softening point of either homopolyimide. This can be seen, for example, in Figure 78, which shows the elasticity modulus as a function of the temperature for the copolymer of dianhydride VI with m-phenylene diamine (R' = 3) and bis-(4-aminophenyl ether)-hydroquinone (R' = 9) taken in equimolecular proportions, and for the corresponding homopolyimides VI-3 and VI-9. The copolyimide softens in a narrower range and at a lower temperature than the homopolyimides (see also Table 31). The homopolyimides VI-3 and VI-9 are capable of crystallizing; this is indicated by the shape of the temperature dependence curve of the elasticity moduli for these polymers (Figure 78,1 and 78,3). The shape of the curve for the copolyimide (Figure 78,2) is, on the contrary, typical of a linear noncrystallizing, softening polymer. It is seen, accordingly, that copolycondensation of polyimides, as in the case of many carbon-chain polymers, is accompanied by a loss in the crystallization capacity, even though each one of the homopolymer pairs has this capacity. If the components have been taken in unequal molecular proportions, the modulus-temperature curve shows two softening ranges near the softening points of the initial homopolymers, as may be seen in Figure 79 (cf. Figure 64).

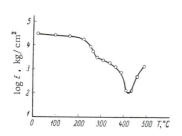

FIGURE 79. Logarithm of elasticity modulus as a function of temperature for the copolymer of dianhydride III with diamines 9 and 10, taken in the ratio 10:9:1.

On the whole, these relationships are also valid for copolymers prepared from one diamine and two dianhydrides.

Table 31 shows the mechanical and thermal parameters of a number of copolymers. It can be seen from these data that copolycondensation may bring about a substantial improvement in the parameters of the final products in certain cases. For example, copolyimide of dianhydrides III and VI with diamine 9 has a better elasticity than either of the two homopolymers III-9 and VI-9 (cf. Tables 26 and 28). At temperatures above the softening point this copolymer becomes highly oriented. For example, if a film of this polymer is stretched by 250% at 280°, the tensile strength increases to 9300 kg/cm², while the high (about 20%) elongation-at-break is retained. It may also be pointed out that many copolyimides display a better adhesion to glass and metals than do the corresponding homopolyimides. It will be seen from these examples that copolycondensation is a promising method for improving the parameters of polyimides.

TABLE 31. Properties of a number of copolyimides

Composition			σ_p, kg/cm²	ε_p, %	T_s, °C	
R	R'	ratio	at 20°C		of copolymer	of homopolymers
VI	3 / 9	1:1:1	1200	20	220	~260 / 250
VI	4 / 5	1:1:1	1400	20	260	310 / 270
III	3 / 5	1:1:1	—	—	270	~300 / >400
III	3 / 9	10:1:9	1600	35	260 / 400	~300 / ~400
III VI	9	1:1:1	1650	80	230	~400 / 250

Orientation stretching and crystallizability of polyimides

It has already been noted that orientation stretching of polyimides is accompanied by an increase in the tensile strength and elasticity modulus. These effects may be very considerable /19/. Thermal stretching of Group A, Group B and Group C polyimides takes place under large stresses and the elasticity moduli are high owing to the absence of a sharp softening point. The limiting elongations are small. Group D polyimides may be stretched above the softening point under small loads to produce high degrees of orientation. Table 32 shows a number of data on the effect of orientation on the mechanical properties of films and fibers made of different polyimides.

Noteworthy features are the high elasticity moduli of Group C oriented polyimides (I-5, I-9) and their high tensile strengths at 400°. In the case of polymer I-9 these effects may also be attributed to crystallization during the orientation, since the stretching is accompanied by a marked increase in density. Crystallization is absent in polymer IX-9 based on the dianhydride of 1,2,3,4-butanetetracarboxylic acid, but the increase in the elasticity modulus as a result of stretching is nevertheless considerable. Thus, the marked increase in the elasticity modulus of stretched polyimides cannot, as a rule, be explained by orientation crystallization alone. It must be realized, however, that when the crystallization does take place, it is obviously accompanied by an increase in the modulus (see, e.g., Figure 72).

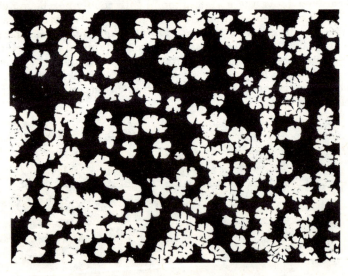

FIGURE 80. Spherulites in film of polyimide VI-9 at 260°C.

Crossed nicols.

When polyamido-acid films are imidized by thermal treatment on a rigid support, the stresses caused by the decrease in volume result in the appearance of a biaxial orientation. These effects are especially marked

TABLE 32. Effect of orientation stretching on mechanical properties of a number of polyimides

Polymer	Nonoriented specimens		Conditions of stretching		Oriented specimens				
	T_s, °C	$E_{20°}$, kg/cm²	temperature, °C	degree of stretching, %	E, kg/cm²	σ_p, kg/cm² (20°)	ε_p, % (20°)	σ_p, kg/cm² (400°)	ε_p, % (400°)
I-2	—	120,000	400	18	130,000	2500	2	—	—
I-5	—	35,000	400	160	80,000	7000	15	1500	20
I-5*	—	35,000	350—400	180	100,000	10000	10	2000	15
I-9*	—	32,000	—	—	300,000—380,000	9000—11000	4—10	1500—5000	5—10
VII-9	200	28,000	280	320	—	7500	9	—	—
IX-9	270	30,000	320	180	70,000	3600	15	—	—

* Data for fiber.

in nonmelting polyimides, for example those in Group C. This orientation also affects the mechanical properties of polyimide films, which acquire tensile strengths and elongations-at-break 30—50% higher than films treated when unsupported /19/. The presence of biaxial orientation was confirmed by measuring the double refringence of cross-sections and by X-ray data /71/.

The crystallinity and capacity for crystallization of polyimides was until recently estimated only qualitatively from the X-ray patterns given by isotropic films. Sroog et al. /101/ (see also Table 5) used the X-ray patterns of polypyromellitimides to arrive at the following conclusions: polymer I-1 crystallizes to a marked extent in the very process of imidization, polymers I-2, I-5 and I-7 crystallize on being heated to 400°, while the extent of crystallization of polymers I-11 and I-12 is low. They found that the degree of ordering attained is not lost after brief heating at 750°. That polymer I-2 is crystalline is indicated by its high density (1.46 g/cm³) and by its exceedingly high elasticity modulus (120,000 kg/cm²) (Tables 21, 24, 32). The evidence for the crystalline nature of polymers I-5 and I-7 is less conclusive, the more so as the X-ray crystallinity of I-5 films proved to be as low as 13% /71/; at such low degrees of crystallinity the X-ray method is no longer reliable.

Smirnova /35/ studied the crystallization capacity of these and other aromatic polyimides using the results of determinations of density and elasticity modulus as well as polarizing microscope data. The crystallizability of polyimides is displayed in certain cases by the formation of spherulites. The spherulite structures shown in Figure 80 are formed when polymer VI-9 is heated at 250—270°, i.e., in its polyimide form. The film then becomes turbid and brittle. On being further heated, the spherulites melt around 450° and the birefringence of the specimen disappears altogether. Spherulites do not reappear when the sample is cooled and reheated. The crystallization is not repeated more than once. If these observations are compared with the data obtained for the elasticity modulus of amorphous (i. e., heated not above 200°) VI-9 film (Figure 81), a full correlation is noted. The decrease in the modulus at about 220° and its renewed rise on further heating indicate that the polymer passes into a highly elastic state and then crystallizes. The decrease in the elasticity modulus at 400° is evidence of melting. On cooling the elasticity modulus increases only in the vitrification range and crystallization effects are not noted. The last-named effect is obviously due to cross-linking. That cross-linking prevents crystallization and fixes the amorphous structure can also be seen in Figure 82, which shows the temperature dependence of the elasticity modulus of a film of this polymer (VI-9) which has crystallized out at 300°. Heating reveals two ranges of sudden change in the modulus, corresponding to softening and melting, while on cooling only one such range is observed. When the heating is repeated, the resulting curve is typical of an amorphous polymer. It should be noted that in the case of VI-9 polymer cross-linking does not really affect the elasticity modulus of the amorphous final product, as distinct from, say, polymer VI-3 (Figure 72a), for which the effect is considerable.

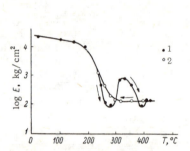

FIGURE 81. Logarithm of elasticity modulus as a function of the temperature for polyimide VI-9.

Film preheated at 200°C (amorphous).
1 — heatup; 2 — cooling.

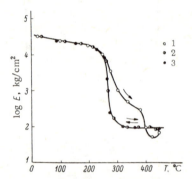

FIGURE 82. Logarithm of elasticity modulus as a function of the temperature for polyimide VI-9.

Film preheated at 300°C (crystallized).
1 — heatup; 2 — cooling; 3 — repeated heatup

If the temporal variation of the elasticity modulus is measured during the thermal treatment of polyamido-acids which are precursors of non-crystallizing polymers, only one zone of sudden changes is noted; this is produced by imidization. In the case of crystallizing polymers there are two such zones — one corresponding to imidization, the other to crystallization. This is shown in Figure 83 which represents the magnitude

$\frac{|E|_{15} - |E|_3}{|E|_{15}}$ (where $|E|_3$ and $|E|_{15}$ are the values of the dynamic elasticity

modulus 3 and 15 minutes after the temperature has been established) as a function of the treatment temperature for films of amorphous (VI-5) and crystallizing (VI-9) polyimides.

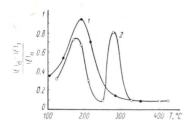

FIGURE 83. Relative temporal increment of elasticity modulus $\dfrac{|E|_{15} - |E|_3}{|E|_{15}}$ as a function of temperature for polyimides VI-5 (1) and VI-9 (2).

Amido-acid films preheated at 80°C for 2 hours.

One of the best and simplest methods for studying phase transformations is the dilatometric method. Direct measurements of volume or density changes at all stages of existence of polyimide, beginning with the polyamido-acid, are difficult owing to the liberation of the water of imidization, presence of residual solvent, high transition temperatures etc. These difficulties may be obviated by quenching the high-temperature structure of the polymer. In this way the density can be measured at room temperature after rapid cooling rather than directly, at a high temperature. It has been experimentally shown that in the case of polyimides it is enough to cool rapidly to room temperature. It is possible to follow in this way the variation in the density of these polymers during stepwise thermal treatment. To do this is a piece of the sample heated at the given temperature and rapidly cooled is cut off and its density determined at 20°, while the remainder of the sample is heated at the next higher temperature, quenched, etc.

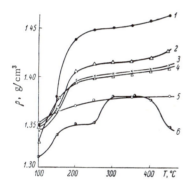

FIGURE 84. Densities of polyimide films measured at 20°C as a function of the treatment temperature.

Initial polyamido-acid films preheated at 80°C for 24 hours. 1 — I-2; 2 — VI-2; 3 — I-5; 4 — I-9; 5 — VI-5; 6 — VI-9.

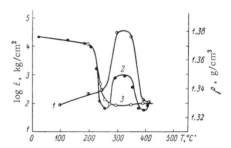

FIGURE 85. Density measured at 20°C as a function of the preheating temperature (1) and logarithm of elasticity modulus as a function of the temperature on heating (2) and cooling (3) of polyimide VII-9.

Initial polyamido-acid films preheated at 40°C for 10 hours.

In these experiments density was measured in a mixture of carbon tetrachloride with toluene by the flotation method. The density thus obtained was assumed to correspond to the most recent heating temperature for this fragment of the sample. The measurements were begun from polyamido-acid films dried at 80°. Figure 84 shows the relationships between the density and the heating temperature, obtained in this manner, for a number of aromatic polyimides of various structures. It is seen that in all cases there was a sudden rise in the density between 100 and 200°. It has been seen that this zone corresponds to intensive imidization. Treatment at temperatures between 200 and 450° does not significantly affect the densities of most polymers, including I-2 and I-5 (which have been reported to crystallize /101/), except for a slight steady increase. In the case of polyimide VI-9 there is a sudden rise in density at 250° from 1.35 to 1.38 g/cm^3; the latter value is retained up to 450°, after which it drops suddenly to the former value (prior to 250°) which then remains constant on cooling and repeated heatup.

The results of density measurements (Figure 84), determinations of elasticity modulus (Figures 81 – 83) and the formation and disappearance of birefringent spherulite structures (Figure 80) noted for polyimide VI-9 when taken together, indicate that we have here in fact crystallization and fusion effects (phase transitions of the first kind). Other polymers, the density of which as a function of the temperature is shown in Figure 84, do not show any such effects. A similar correlation between the variation of density and elasticity modulus with the temperature for another crystallizing polyimide (VII-9) is illustrated in Figure 85.

As in the case of any other polymers, the crystallization of polyimides is facilitated in the presence of solvents. Figure 86 shows the variation in density during thermal treatment of two samples of polymer VI-9. The conditions of the treatment (temperature and holding times) were the same in both cases, but one of the samples was held for some time in dimethylformamide (solvent) vapor at 150°; the density of this sample immediately showed a considerable rise and attained 1.38 g/cm^3 (Figure 86, curve 1) which is the maximum value for polyimide VI-9. The sample which had been heated at all temperatures in the absence of the solvent vapor, assumed this density value only after treatment at 300° (curve 2). It is the general rule for polymers that a sample which crystallized at a lower temperature also melts at a lower temperature, as may be seen from the descending parts of the curves in Figure 86 between 400 and 450°.

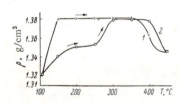

FIGURE 86. Density of VI-9 polyimide film measured at 20°C as a function of the heating temperature.

1 — heated at 150°C in dimethylforma- mide vapor; 2 — heated in vacuo.

It should be noted that the treatment in the solvent vapor at 150° resulted in the crystallization of the already imidized polymer, as indicated by its insolubility. Most polyimides may be crystallized in the presence of solvents.

It may be concluded /35/ that crystallization resulting from heating under conventional conditions, which is accompanied by distinct phase transitions of the first kind, can occur only in polyimides which soften and whose structure places them in Group D. This is a necessary, but not always a sufficient condition. In fact, as has just been shown, polyimides VI-9 and VII-9 in this group, which are based on 1,4-bis-(4-aminophenoxy) benzene, crystallize very readily. Judging by the sudden density changes with the conditions of preparation of polyimides III-10, V-4, VI-3 (Table 21), these polyimides would also appear to be of the crystallizing type. In all these cases crystallization takes place once only; the cross-linking which takes place in the molten state prevents a crystalline structure from being formed on cooling. On the other hand, polyimides in Group D such as VI-5, VII-5 or VI-7, which very readily soften and which display only very limited crosslinking, show no signs of crystallization. The dependence between the crystallization capacity and the structure of the repeating monomer unit is complicated. We may merely note that the density data in Table 21 seem to indicate that the introduction of side groups into R', replacement of ether by sulfide bonds and the presence of aliphatic or alicyclic radical in the dianhydride component render crystallization more difficult.

As regards the nonmelting, nonsoftening polyimides belonging to Groups A, B and C, it may be said that since there is no clear evidence of phase transitions of the first kind in a wide range of temperatures, the problem of crystallization of these polymers requires further study. In any case, thermal treatment of isotropic films of these polymers is not accompanied by any conspicuous effects indicating crystallization (Figure 84).

All that has so far been said does not mean, however, that nonsoftening polyimides cannot in special cases be obtained in a highly crystalline state. This may probably be achieved by crystallization in a swollen state or in solution, by orienting the polyamido-acid and transferring the oriented structure to the polyimide, by crystallization at the amido-acid stage, etc. In particular, the high density values of polyimides I-2, I-8 and VI-1 (Table 21) may indicate crystallization on passing from the polyamido-acid to the polyimide. The high values of the elasticity moduli of oriented fibers of I-5 and I-9 (Table 32) may indicate orientation-produced crystallization, etc.

Certain optical properties of aromatic polyimides

Many of the characteristic properties of polyimides are undoubtedly directly connected with the physical characteristics of polyimide macromolecules. The study of the properties of these macromolecules (and also of other ring-chain macromolecules) is seriously impeded by the insolubility of most polyimides and by the necessity of allowing for the multi-step nature of their preparation. One method of studying the structure of macromolecules in polymers in the bulk is the determination of double refraction under mechanical stress — the study of the photoelastic effect. The theory of photoelasticity of polymeric networks in a resin-like state makes it possible to calculate the optical anisotropy of the macromolecular segment and estimate the flexibility of the macromolecules from the experimental data. The temperature variation of the photoelastic coefficient makes it possible to identify transitions of the softening and melting type and to take qualitative note of the chemical transformations undergone by the polymer.

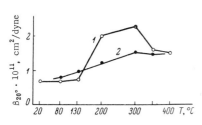

FIGURE 87. Photoelastic coefficient (measured at room temperature) during thermal treatment of PM films (1) and DFO films (2) with 30 minutes' holding at each treatment temperature, beginning from 80°C.

113

Such experiments were performed by the authors of this book /4/ on two polyimides with very different thermomechanical properties: PM (I-5) and DFO (VI-5). The corresponding polyamido-acids were also studied. The birefringence Δn and the deformation Δl were determined at various temperatures between 20 and 400°, under several small stretch loads P in the range of reversible, directly proportional dependence of Δn and Δl on P. The data thus obtained served to calculate the photoelastic coefficients for the resin-like state

$$\beta = \frac{\Delta n \cdot S_0}{P}$$

and the elasticity modulus

$$E = \frac{P \cdot l_0}{S_0 \cdot \Delta l},$$

where S_0 and l_0 are the cross-sectional area and the length of sample, respectively.

The photoelastic coefficient of polyimides at room temperature β_{20} proved to be one or even two orders of magnitude greater than that of low-molecular glasses and solid linear chain polymers: $100 \cdot 10^{-13}$ to $200 \cdot 10^{-13}$ as against $1 \cdot 10^{-13}$ to $10 \cdot 10^{-13}$ cm^2/dyne. In addition, it displays a considerable increase when the chain structure changes — when polyamido-acid is converted to the polyimide which is known to have a markedly higher polarization anisotropy of the monomeric unit (Figure 87). Solid state deformation, even in the purely elastic range, produces not only changes in inter-atomic distances, but also rotations of individual large segments of macromolecules. Unless this were so, there would be no major difference in the β_{20} values between polyimides and linear chain polymers which have small proper anisotropy and changes in β_{20} produced by imidization. It is seen from Figure 87 that measurements of the photoelastic coefficient β_{20} can be utilized to follow the course of imidization and probably also of cross-linking produced by ring rupture. In any case, this conclusion is not in contradiction with the drop in β_{20} after heating at 350° in the case of PM polymer (strong cross-linking) and the absence of such drop in the case of DFO polymer (weak cross-linking).

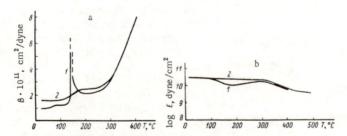

FIGURE 88. Temperature dependence of photoelastic coefficient (a) and elasticity modulus (b) for polyamido-acid (1) and polyimide (2) PM.

The course of temperature variation of coefficient β and elasticity modulus E for the polyamido-acid and polyimide PM (Figure 88) indicates that the polyamido-acid softens around 150° but the softening process is superposed by the process of imidization. As a result the values of β and E of both polymers are similar above 200–250°. The same applies to polyimide DFO (Figure 89).

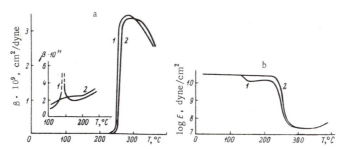

FIGURE 89. Temperature dependence of photoelastic coefficient (a) and elasticity modulus (b) for polyamido-acid (1) and polyimide (2) DFO.

At high temperatures the polyimide DFO passes into a highly elastic state. Between 250 and 270° the modulus of elasticity decreases by three orders of magnitude, while the coefficient β increases by more than two orders. In accordance with the kinetic theory of elasticity and bire-fringence of polymeric lattices above 270°, the quantitites βT and $E\frac{1}{T}$ remain constant /4/ and the sign of β remains positive. This is a direct consequence of the presence of ring structures in the main chain of the macromolecule.

According to the theory of birefringence of deformed polymeric lattices /37/, the coefficient of photoelasticity β for uniaxial extensions is connected with the intrinsic anisotropy $(a_1 - a_2)$ of the chain segment by the relationship

$$\beta = \frac{2\pi}{45kT} \cdot \frac{(n^2 + 2)^2}{n} \cdot (a_1 - a_2),$$

where n is the average refractive index, T is the absolute temperature and a_1 and a_2 are the tensor components of the polarizability of the segment along its axis and the average value of the component across the axis, respectively.

Data in Figure 89 may be used to calculate the anisotropy of segments of the macromolecules of DFO polyimide with the aid of the above expression. The flexibility of macromolecules is usually characterized as the number s of monomeric links in one chain segment. The magnitude s connects the anisotropy of the segment with the anisotropy of the mono-meric unit by the expression

$$(a_1 - a_2) = s(a_\| - a_\perp),$$

where $a_\|$ and $a_\perp$ are the respective polarizabilities along and transverse to the axis of the monomeric unit. The latter may be estimated from the geometric model of the unit, using the polarizability data of the unit-constituting links and the principle of additivity of the components of the polarizability. This has been done for the flat model of monomeric unit of DFO, both assuming a rigid structure and assuming a fully free rotation around $C_{ar}-O$ and $C_{ar}-N$ bonds. The results of these calculations are shown in Table 33, together with the literature data for polystyrene.

TABLE 33. Results of studies of photoelasticity and estimates of chain flexibilities /4/

Polymer	B, cm^2/dyne		$(\alpha_1-\alpha_2) \times$ $\times 10^{25}$, cm^3	$(\alpha_{||}-\alpha_{\perp}) \times$ $\times 10^{25}$, cm^3	s	$E \cdot B \cdot 10^4$ $(T>T_S)$
	20°C	$T > T_S$				
Polystyrene	$+9 \cdot 10^{-13}$	$-4 \cdot 10^{-10}$	-126	-18	7	6
Polyimide DFO	$+120 \cdot 10^{-13}$	$+33 \cdot 10^{-10}$	$+1350$	$+210$	6.5	800

It is seen that the anisotropy of the segment and of the DFO repeating unit is opposite in sign and dozens of times greater in absolute magnitude than the corresponding parameters of polystyrene molecules, in which the most anisotropic groups — phenyl nuclei — are located in side grafts. However, the number of units in the segment and thus also the flexibility of the chain are equal in both cases (it should be remembered, however, that in DFO each unit executing free rotation around a bond is larger). Clearly, the actual sizes of the segment and the unit are much larger in DFO than in polystyrene.

The last column of Table 33 shows the values of the product of elasticity modulus and the photoelastic coefficient. The magnitude is a characteristic of the quality of the material on which mechanical stresses are studied by the photoelastic method. It is seen that this quality in aromatic polyimides may be very high, and the polymers themselves may be potentially suitable for such uses.

A simple and sensitive method of study of the molecular structure is the refractometric method /15/. A comparison between the experimentally determined molar refraction with that calculated from bond refractions makes it possible to arrive at certain conclusions regarding the structural formula of the molecules and some of their structural features (conjugation effects, mutual effects of bonds etc.). The experimental values of molar refraction R are usually obtained by applying the Lorentz and Lorenz equation

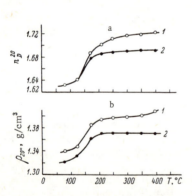

FIGURE 90. Variation of refractive index (a) and density (b) in the process of thermal treatment for polyimides PM (1) and DFO (2).

Films predried at 23°C for 24 hours and at 80°C for 1½ hours. They were held for 15 minutes at each succeeding temperature.

$$R = \frac{n^2-1}{n^2+2} \cdot \frac{M}{\rho},$$

where n is the average refractive index, M is the molecular weight of the substance and ρ is its density. For an additive mixture of substances we have $R = \Sigma x_i R_i$ where x_i and R_i are the molar fraction and molar refraction of the i-th substance, respectively. It has been shown that these relationships are applicable to linear carbon-chain polymers.

The authors of this book determined the refractive indexes n_D^{20} and densities ρ_{20} at room temperature for the following aromatic polyimides: BP (I-2), PM (I-5), DBF (III-2) and DFO (VI-5).

These parameters were determined after individual stages of thermal treatment of the polyamido-acids. The refractive index was determined with the aid of an Abbé-refractometer and by the immersion method, while the density was determined by the flotation method. All measurements were performed on films.

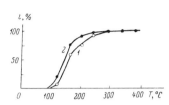

FIGURE 91. Degrees of imidization i at different stages of thermal treatment of polymers PM (1) and DFO (2) calculated from the results of refractometric measurements.

Typical results obtained for n_D^{20} and ρ_{20} for two of the polyimides are shown in Figure 90. It is seen that a rapid rise in density and refractive index is encountered in the temperature range of intensive imidization $(130-200°)$. Assuming that the low-temperature values of n_D^{20} and ρ_{20} are those of pure polyamido-acid, while the high-temperature values refer to pure polyimide, it is possible to calculate the molar refractions and the degree of imidization at different stages of thermal treatment (Figure 91). It is seen that the rate of imidization and its temperature range are the same in both cases.

Experimental and calculated values of refractive parameters of the polyamido-acids and polyimides are shown in Table 34.

TABLE 34. Results of refractometric measurements effected on polyamido-acids and polyimides

Property	PB (I-2)		DFB (III-2)		PM (I-5)		DFO (VI-5)	
	acid	imide	acid	imide	acid	imide	acid	imide
Density, ρ_{20}, g/cm^3 ..	1.333	1.431	1.33	1.40	1.340	1.410	1.322	1.370
Refractive index, n_D^{20} ..	1.68	1.78	1.70	1.78	1.632	1.722	1.630	1.690
Molecular weight of repeating unit	402	366	478	442	418	382	510	474
R_{exp}	114.0	108	139	132	111.3	107.0	137.1	132.2
R_{calc}	103.2	97.1	127.3	121.2	104.9	98.8	130.6	124.5
$\Delta R = R_{exp} - R_{calc}$ ...	+11.3	+10.9	+11.7	+10.8	+6.4	+8.2	+6.5	+7.7

The high values of the refractive index are noteworthy. Those of carbon chain polymers usually do not exceed $1.5-1.6$ and for some types of phenol-formaldehyde resins they are as high as 1.7 /22/. For fully aromatic polyimides PB (I-2) and DFB (III-2) the value of the refractive index is higher than 1.78.

According to a rough estimate, n_D^{20} for PB and DFB should be about 1.82. Such high refractive indexes are displayed only by special optical glasses /22/. For polyimides with ether bonds in the chain $n_D^{20} \geqslant 1.69$.

The measured values of molar refraction R_{exp} for both polyamido-acids and polyimides are invariably higher than the values R_{calc} calculated from atomic and group refractions of Vogel /15/ or from any other literature data. The large positive exaltation value ΔR indicates that the calculation leaves out of account some inter-atomic and inter-bond effects in the ring-chain polymeric molecule. A comparison of the exaltation values for purely aromatic polyimides PB and DFB with those of oxygenated polymers PM and DFO clearly shows that in the former case the inter-bond (conjugation) effect is much stronger and that the introduction of hetero-atoms into the chain reduces this effect.

Chapter IV

PRACTICAL UTILIZATION OF POLYIMIDES

It.is now known that practically all types of technical materials in which polymers are used in the solid state can be produced from polyimides. Electroinsulating film, enamel insulation of coil ducts, sealing compounds and glues are now produced from polyimides on a semi-industrial scale. Preparation of fibers and plastic materials is being developed. The intermediate products — diamines and dianhydrides — have found use as hardeners of epoxide resins, which considerably raise their working temperatures; dianhydrides of pyromellitic and cyclopentanetetracarboxylic acids /52, 54, 106/ have been found to be particularly promising.

Detailed information on the laboratory methods of preparation of polyimide materials and their exact chemical composition can be found mainly in patent literature. The properties of industrial polyimide materials, except for information about their composition or production methods, are described in scientific periodicals.

Intensive work in this field is conducted mostly in the U.S., in the Soviet Union, Japan, West Germany and other countries. The technological and commercial importance of polyimides is responsible for the competition between large American corporations such as DuPont, General Electric, Westinghouse Electric, Amoco Chemicals, Monsanto, Narmco and others. According to the data published by DuPont, the production of these polymers in the U.S. will attain 10,000 tons/year in 1970, as compared with 100 tons per year in 1964 /55/. Aromatic polyimides are industrially the most important.

Polyimide films

Experimental production of polyimide film was begun by DuPont in 1962. This organization now manufactures two types of film under a common trade name of "Kapton": "H-film" made of pure polyimide, and "HF-film" made of this polyimide and coated with polytetrafluoroethylene on one or both sides. The thickness of the film is between 12.5 and 150μ, and its price is about 25 dollars per pound /83/. The leaves are transparent and are golden-yellow when thin and brown when thick. The properties of these films have been described in a number of papers published by employees of DuPont and by organizations which carry out routine testing of the samples and also in many advertisements /51, 62, 82, 84, 90, 108, etc./.

It may be deduced /47/ from the IR spectra and physical and mechanical properties of "H-film" that it consists of a polyimide of pyromellitic acid dianhydride and 4,4'-diaminodiphenyl ether (PM).* Commercial "H-film" does not seem to be fully imidized. Judging by the analysis of its thermal degradation products, its degree of imidization is approximately 0.8/47-49/.

"H-film" has a very interesting ensemble of physical and mechanical properties, and almost none of its important technological parameters especially those at high and low temperatures, has been duplicated by any other known polymeric film materials.

The mechanical properties of "H-film" usually remain at the ordinary quality specification level throughout the temperature range between liquid helium and very high positive temperatures. For example, at 500° "H-film" is twice as strong as polyethylene film at room temperature (350 against 175 kg/cm^2). Its strength at room temperature is equal to that of poly(ethylene terephthalate) film, while at below-zero temperatures it is much higher. "H-film" does not soften, does not melt and has high values of elasticity modulus — such as are typical of solid polymers — up to 300 — 400°. It is nevertheless extremely flexible; between —200 and 400° the elongation-at-break is not less than 30%. The film retains some flexibility (does not break on being bent around a thin rod) even at —269° /38, 104, 107/. The temperature variations of the major mechanical parameters of "H-film" are shown in Figure 92.

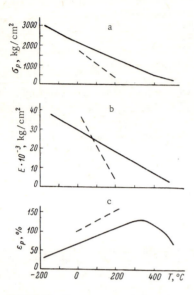

FIGURE 92. Temperature variations of tensile strength (a), elasticity modulus (b) and elongation-at-break (c) of polyimide "H-film" /104/ and polyester film "Mylar" (poly(ethylene terephthalate)) /38/.

Continuous line — "H-film"; broken line — "Mylar."

Under certain conditions "H-film" is even more temperature-resistant than metals. Thus, its so-called zero strength temperature** is 815°, which is 300° higher than this parameter for aluminum. The film is not decomposed even when briefly held in the jet of a plasma torch.

As regards its dielectric properties, "H-film" is a weakly polar, medium-frequency dielectric of polycarbonate and polyester type. At room temperature the main dielectric characteristics of "H-film" are about the same or slightly superior to those of commonly used polyester films (Lavsan, Terylene, Mylar). In the case of "H-film," however, these properties are much less dependent on the temperature (Figure 93). For this reason the limiting working temperatures of this material when used as dielectric are much higher. For example, for medium frequency

* For meaning of these conventional designations see Chapter III, Table 19.
** A test commonly accepted in the U.S. A film or foil sample under a small tensile load (about 1.5 kg/cm^2) is brought into contact with a heated rod. The temperature of the rod which produces disintegration within 5 seconds is considered to be "zero strength temperature."

condenser dielectrics the permissible value of the tangent of dielectric loss angle is tan $\delta \leqslant 0.005$, while that of specific volume resistance is $\rho_{v} \geqslant 10^{13}$ ohm · cm. It is seen from Figures 93b and 93c that "H-film" satisfies these specifications up to at least 250°. The permissible working temperatures for polycarbonate and polyester films are only 160—180 and 100—120°, respectively. The main requirements for low-frequency dielectrics are high resistivity ($\rho_{v} \geqslant 10^{11}$ ohm · cm) and electric strength. "H-film" can be used in this capacity at 300° and at higher temperatures (Figures 93c and 94).

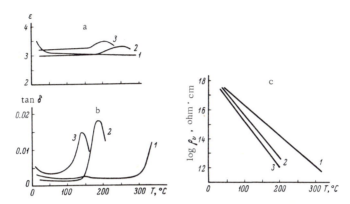

FIGURE 93. Temperature variation of dielectric constant (a), tangent of dielectric loss angle (b) at 10^{3} hz and specific volume resistance (c) of "H-film" (1), polycarbonate film (2) and polyester film (3) /17, 38, 69, 104/.

It is important to note that the polyimide film does not merely display superior mechanical and electrical parameters at high temperatures, but also retains them for a long time under these conditions. Very instructive in this respect are the results of studies of the thermostability of "H-film" in which its mechanical parameters were determined at room temperature, after aging in the air and in inert media at high temperatures /69/. It was found that the temporal dependence of the tensile strength σ_{p}, impact strength σ_{i} and relative elongation-at-break ε_{p} may be described by a simple empirical equation of the type

$$\log p = \log P_{0} - kt,$$

where P_{0} and P are the initial and the current values of the mechanical parameters, respectively, t is the time and k is a constant which describes the rate of variation of the parameter at the given temperature. It is seen from Figure 95, which represents the variation of the elongation-at-break after aging in the air at 300°, that this relationship is in good agreement with the experimental data. The variation of the constants k with the temperature obeys Arrhenius' equation:

$$k = k_{0}e^{-\frac{U}{RT}}.$$

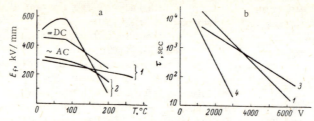

FIGURE 94. Electric strength as a function of temperature for DC and AC (a) and time to breakdown as a function of the voltage at 20°C for AC (b) for different polymer films /38,69,104/.

1, 2, 3 — as Figure 93; 4 — Teflon film.

If k is plotted as a function of reciprocal temperature (Figure 96), k_0 and U can be calculated. The numerical values of the coefficients are given in Table 35.

TABLE 35. Numerical parameters of variation of mechanical properties of "H-film" during aging /69/

Parameter	P_0	U, kcal/mole		$\log k_0$, hr^{-1}		$k_{400°}$, hr^{-1}	
		air	helium	air	helium	air	helium
ε_p, %	70	40	54.5	12.0	15.3	0.15	0.005
σ_p, kg/cm^2	1600	39	55.5	11.2	15.2	0.04	0.002
σ_i, kg·cm/cm	2400	41.5	51.0	12.5	14.0	0.14	0.003

N o t e .   Impact strength σ_i was determined from the energy loss of a falling sphere at the moment of film rupture.

Using the data in the table and bearing in mind the known limiting permissible values of mechanical parameters, the duration of service can be fairly reliably determined. Examples of experimental duration of service of "H-film" and service times calculated by extrapolation in the air and in helium at different temperatures are shown in Table 36.

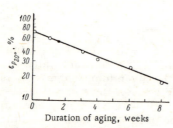

FIGURE 95. Relative elongation-at-break, measured at 20°C, as a function of time of aging in the air at 300°C for "H-film" /69/.

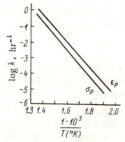

FIGURE 96. Logarithms of rate variation constants of tensile strength and elongation resulting from aging in the air as a function of reciprocal absolute temperature for "H-film" / 69/.

It was assumed that the duration of service is the time during which the relative elongation decreases to $\varepsilon_p = 1\%$. In an inert atmosphere the service times were $5-10$ times higher than in the air. It is seen from Table 36 that "H-film" is much more thermostable than poly(ethylene terephthalate) and Teflon films.

The thermal stability of the electric parameters of "H-film" is even higher than that of mechanical parameters. After aging for 8 weeks in the air at 300°, when the mechanical parameters of the film have deteriorated to a fraction of their former value, the breakdown electric strength decreases by only 20%, while the dielectric constant and the tangent of dielectric loss angle increase by 10%. The specific volume resistance increases considerably (Figure 97).

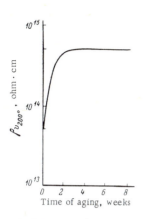

FIGURE 97. Specific volume resistance (measured at 200°C) as a function of the time of aging of "H-film" in the air at 300°C /69/.

Such data indicate that polypyromellitimide film gives protracted service up to $250-275°$ in the air and up to $325-350°$ in an inert medium. Its performance is superior to that of any other known polymeric film material.

The stability to radiation of "H-film" is also exceptionally high /101/. A film exposed to a dose of 10,000 Mrad of fast electrons retains its flexibility and its electric parameters remain unchanged. These parameters also remain unchanged after prolonged irradiation with thermal neutrons (dose about $5 \cdot 10^{18}$ neutrons/cm^2); it may be mentioned in this connection that even polystyrene, which is the most highly radiation-resistant of all known polymers, is fully decomposed when irradiated with doses $10-15$ times smaller.

"H-film" does not dissolve or swell in organic solvents and oils, is resistant to dilute acids, but is decomposed by concentrated alkalis and superheated steam. Moisture absorption is relatively high — 1.3% in 24 hours at 20° and 50% relative humidity.

Another very interesting product is the "HF-film," which consists of a polyimide film coated with one or two layers of Teflon (on one or on both sides). As regards the mechanical and electric properties at high temperatures, it is somewhat inferior to "H-film," but has a higher chemical stability, resistance to moisture, is arc-resistant and can be welded if heated to the melting point of Teflon.

The main physical and mechanical parameters of "H-film" and "HF-film" are tabulated in Table 36.

It follows from the discussion of the general physical properties of polyimides, given in Chapter III, that the polypyromellitimide PM, which is the base of "H-film" is not the only polymer of this class suitable for the preparation of films. The polyimide PFG, which is stronger and more elastic when oriented, is just as suitable. Films made of the polyimide DFO are also very strong and elastic and, unlike PM and PFG films, are capable of softening. A thermoplastic film with a high softening point may be prepared from DFFG polyimide. It will thus be seen that this polymer class includes highly promising starting materials for the manufacture of thermostable films of different properties and intended for various purposes.

TABLE 36. Physical and mechanical properties of commercial films 25 — 35 μ thick /38, 69, 82, 90, 107/

Brand name	Density at 20°C, g/cm³	Softening temperature, °C	Melting point, °C	Zero strength temperature, °C	Tensile strength, kg/cm²		Elongation-at-break, %		Elasticity modulus, kg/cm²	
					20°	200°	20°	200°	20°	200°
Mylar	1.39	70	260	230 — 240	1600	350 — 500	100	125	35000	3500
Teflon	2.15	85	290	250 — 260	210	14	300	175	3100	140
H-film	1.42	None	None	815	1600 — 1800	900 — 1200	70	90	29000	18000
HF-film	—	None	None	—	1100	700	75	75	—	—

TABLE 36 (continued)

Brand name	Impact strength at 20°C, kg·cm/cm	Service time in the air and in helium (in parentheses)						Dielectric constant at 20°C and 10³ hz	Volume resistivity, ohm·cm		Electric strength at 20°C, kV/mm
		200°	250°	275°	300°	350°	400°		20°	200°	
Mylar	2400	1.5 months	Melts	—	—	—	—	3.2	10^{18}	$5 \cdot 10^{11}$	275
Teflon	1600	—	—	Melts	—	—	—	2.1	10^{17}	10^{17}	160 — 200
H-film	2400	—	8 — 10 years	1.5 years	3 months	6 days (1 year)	12 hours (14 days)	3.5	10^{18}	$10^{13} - 10^{14}$	275
HF-film	—	—	—	—	—	—	—	3.0	—	—	180

At present polyimide films are mainly employed as heat-resistant lining and coil insulation of generators of class "H" (180°) and higher classes, and also of electric cables. Polyimide films used for such purposes have not only a high thermal stability, but also a high strength, flexibility and resistance to perforation at high temperatures, which is attained at insulation thicknesses much lower than is usual. Technological operations on polyimide films are performed on conventional machinery. Thus, for instance,/82/ round and rectangular copper cores of electric cables can be wound in this manner. If "HF-film" is used for the purpose and the insulated cable is heated in an induction or ordinary electric furnace at $350-400°$, a monolithic electroinsulation and water-proofing of high quality is obtained. A two-layer winding insulation of "HF-film" of a total thickness of 180μ ensures the proper performance of the cable at 15 kV. This is accompanied by a considerable increase in the maximum permissible working temperature, i.e., the transmitted power, whereas the cable weight is $35-50\%$ lower than that of ordinary cables. The elasticity, strength and satisfactory adhesion of the insulating material to the metal after baking makes it possible to bend thick cores at acute angles, which is particularly important in manufacturing the windings of large size generators.

Of special interest is the utilization of polyimide films in flexible printing circuits /82/. Polyamido-acids are brought onto an oxidized copper foil and heat-treated together with it. The copper is then etched in the desired places and the different design elements applied by heat-sealing or in another manner. A ready-made "H-film" may be glued onto the copper for the purpose. If a one-side "HF-film" is employed, it is pressed onto the copper foil at 280°. A two-side "HF-film" will produce multi-layered printed circuits by pressing several single designs into a block.

Thin polyimide films may be utilized as condenser insulation at working temperatures up to 250° (this cannot be achieved by any of the other known organic condenser dielectrics). This was found by direct testing of condenser sections with polypyromellitimide film $8-25\mu$ thick used as dielectric. According to /38/, the resistance of the insulation of three-layered sections of $\sim 0.05\mu$F capacity with "H-film" about 10μ thick was $10^{13}-10^{14}$ ohms at room temperature and at +200° it was one order higher than that of polyester and two orders higher than that of polycarbonate condenser sections; the tangent of loss angle between -60 and $+250°$ did not exceed $5 \cdot 10^{-3}$, and the temperature coefficient of capacity was $-5 \cdot 10^{-4}$ degrees $^{-1}$. The sections withstood prolonged (more than 1000 hours) testing at 200 and 250° in the air under a voltage of 25 kV/mm without breakdown and without a change in the electric parameters. The advantages of using polyimide condenser insulation can also be perceived on inspecting Figure 98 which represents the time constants of polyimide, poly(ethylene terephthalate), polystyrene and paper condenser sections at different temperatures and the variation of their capacities with the temperature. It is seen that the time constants of polyimide sections are highest at all temperatures. The capacity of polyimide sections varies with the temperature in about the same manner as in the case of polystyrenes.

In modern condenser manufacture special efforts are made to reduce the size (specific volume) of the condensers. Best results have been obtained with electrolytic condensers in which the specific volume attains $0.2\,\mathrm{cm}^3/\mu\mathrm{F}$.

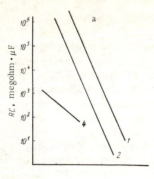

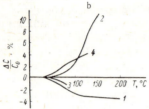

FIGURE 98. Time constant RC (a) and capacity (b) as a function of temperature for condenser sections ($C = 0.06 \, \mu F$) for different dielectrics /38/.

1 — "H-film"; 2 — Terylene; 3 — polystyrene; 4 — paper.

This type of condenser cannot, however, be utilized at high temperatures. Varnish film condensers are known /29/ in which the dielectric is represented by thin $(1-3\mu)$ polymer layers applied from the lacquer solution, while the plates consist of a thin metal layer applied by spray-coating. These condensers are small in size, but in actual fact only monolayer condensers of this type have been manufactured so far, since it is very difficult to select polymer solutions from which subsequent dielectric layers can be applied without swelling and disintegration of the preceding layers. This difficulty was overcome by the authors of this book /33/ who used polypyromellitimide PM as the dielectric. They took advantage of the fact that the cured polyimide layer is not decomposed by polyamido-acid lacquer solution and is readily metallized by thermal spraying in vacuo. Thus, the metal and dielectric layers could be applied several times in alternation and film condenser sections obtained which had a capacity of about $0.1 \, \mu F$ and a size of $(18 \times 2.5 \times 5) \cdot 10^{-3} \, cm^3 = 0.2 \, cm^3$. The assembled condenser models had a specific volume of about $1 \, cm^3 / \mu F$; the tangent of dielectric loss angle at 50 khz between 20 and 300° was about 0.01, and the breakdown voltage was about 20 V. Such condensers are undoubtedly promising.

The use of polyimide films as construction materials proper has not been studied to any extent, but the films can be so utilized owing to their flexibility and resistance to dynamic fatigue loads. The use of "H-film" as membrane in a fuel pump has been reported /107/.

Polyimide bulk plastics

The preparation of bulk plastics from polyimides, in which the advantages offered by these polymers are fully realized, is technologically more difficult than the preparation of films. The difficulties are due, most of all, to the necessity for removing large quantities of solvents (solutions of polyamido-acids usually contain not more than $20-30\%$ dry matter) and water of imidization. When thick layers of polyamido-acids are heated, the imidization water is slowly removed from the reaction zone, which results in a hydrolytic degradation of the macromolecules. Direct conversion of the concentrated solution of a polyamido-acid into a polyimide block, as is the practice in the case of epoxide resins, has not been realized. Accordingly, plastics from polyimides are prepared by isolating the polyamido-acid from solution as thin films, powders, coatings on glass cloth, etc., and converting them fully or partly to polyimides by thermal or chemical methods; thereafter bulk objects are made by pressing, baking, or by some other technique.

The difficulties involved in each of these operations will depend on the type of polyimide. They are very considerable for highly thermostable polyimides such as polypyromellitimides which do not soften to a significant extent, especially at temperatures corresponding to intense cross-linking. The processing may often be further complicated by crystallization. Poly-imides with a sharp transition to highly elastic and viscofluid state, and in which the cross-linking occurs mainly at temperatures above the transition point, may be worked into bulk objects without special difficulty. The preparation of plastics from polymers of this type (Group D), which contain $-O-$, $-CO-$, $-S-$, $-SO_2-$, etc. groups as "hinge" groups in both components of the monomeric link and also meta-bound phenyl nuclei, has been described in patent literature and tested by the authors of this book. Thus, for instance, interesting results were obtained /34/ by studying the processing of polyimide DFO (VI-5) which contains $-O-$ ether groups in both the diamine and the dianhydride components of the monomeric unit. This polyimide softens at 270°, becomes highly cross-linked above 400° and is thermoplastic between 300—400°, i.e., is in a viscofluid state. Its viscosity at 380° is about 10^4 poises, which means that it can be not only pressed, but also pressure-molded. DFO could thus be worked into discs and different electrical engineering objects, both unfilled and filled with glass cloth or with finely dispersed inorganic powders.

In the former case the molding composition consisted of thin (30—50μ) DFO films cut into small pieces. Polyamido-acid films were cast on glass, dried at 80°, lifted off the glass and finally dried and imidized in vacuo at 130, 150, 180, 200 and 250° for 30 minutes each and at 300° for one hour. Powder-filled material was prepared by precipitating the polyamido-acid from solution in dimethylformamide by a mixture of benzene with acetone as a mass of loose consistency. The mass was dried in vacuo for one hour at 100°, heated at 160° and then ground to fine powder. The powder was then imidized in vacuo under the same conditions as the films, ground again and used for the preparation of the press composition. Glass-filled composition was obtained by impregnating a washed and ignited glass cloth ribbon with 15% solution of the polyamido-acid under pressure; this was followed by drying and thermal treatment according to the same method.

Direct pressing was carried out at 370 — 390° under 500 — 2000 atm. pressure, depending on the structure of the object. The press-mold was cooled under pressure. When casting in a press mold, the mold tempera-ture was 370°, the temperature of the feed chamber was 380°; one batch of material served to prepare up to ten objects in the shape of rings with fine radial apertures. These experiments showed that DFO (and several other Group D polyimides) may be worked as ordinary thermoplastic resins. Figure 99 shows photographs of articles made of polyimide plastics obtained by pressing and press-molding unfilled material and by pressing glass-filled material.

Testing films and plastics made of DFO showed that this polyimide has physical and mechanical properties which are superior to those displayed by the thermoplastic resins now on the market (Table 37). Its performance is particularly satisfactory as regards temperature resistance and thermal stability. The Vicat thermostability of DFO is 270°, which is 80° higher

TABLE 37. Physical and mechanical properties of polyimides and other thermoplastic resins

Material	Density, g/cm³	Elasticity modulus, kg/cm² 20°C	Tensile strength, kg/cm²			Elongation-at-break, %			Vicat T_s, °C	Long service temperature, °C	Impact strength, kg·cm/cm³	Moisture absorption in water during 24 hours, %
			−150°	+20°	+200°	−150°	+20°	+200°				
DFO(VI-5), unfilled /34/	1.38	32,000 (12,000 at 200°C)	2500	1400	800	40	70	120	270	200 — 220	50 — 80	0.8
DFO + 40% glass cloth	1.90	100,000		3200 (flexural strength) (nonoriented film)			(nonoriented film)		—	—	160 — 240	—
DFM (VI-3), unfilled	1.40	30,000 (10,000 at 400°C)	—	—	—	—	—	—	490	250 — 300	100 — 200 (filled)	—
Poly(phenylene oxide) /63/	1.07	26,000	—	750	350 at 125°	—	50 — 70	80 at 125°	190	110	—	—
Nylon 6-6 /63/	1.15	12,500	—	800	—	—	150 — 200	—	230	80 — 100	—	7 — 9
Polycarbonate /17,63/	1.2	24,000	—	600 — 800	—	—	—	—	135	—	—	—
Poly(methyl methacrylate)	1.1	27,000	1200	700	—	2 — 3	10	—	110	60 — 80	20 — 25	—

than that of the most heat-resistant thermoplastic resin known — poly
(ethylene oxide) /63/. For this reason DFO retains a high strength and
rigidity at 200° and above, when other thermoplastic resins yield to even
very small loads.

FIGURE 99. Articles made of polyimide plastics.

The extent of thermostability of DFO is illustrated by the fact that its
technological parameters are preserved on being held in a mold at 390°
for one hour and that it withstands repeated pressing. DFO films held in the
air at 380° for one hour retain practically their entire tensile strength
and elongation-at-break. When held for three hours, their strength
decreases by 10%, and their elongation-at-break by 40—50% (Table 29).
When held in the air at 250° for 500 hours, the strength and elongation of
DFO films decreases by not more than 10% of the initial value. The
thermostability of DFO is superior to that of many other Group D polyimides which are suitable for the preparation of thermoplastic materials.
The reason for it is, obviously, that the ether bond between the phenyl
radicals in both polymer chains and in model compounds /64/ is more
stable to heat than any other flexible bonds.

DFO polymer is highly stable to UV light. The mechanical properties
of DFO film did not deteriorate after 200 hours' irradiation with PRK-2
mercury lamp, while Lavsan, polyethylene and capron films tested for
comparison cracked and totally lost their elastic properties. The mechanical and thermomechanical properties of DFO samples were not
significantly affected, and their softening properties were unchanged by
the following doses of penetrating radiation, which had a destructive
effect on other polymers: γ-rays ($\sim$ 10,000 Mrad) (Table 38), fast electrons
($\sim$ 10,000 Mrad) and neutrons ($\sim 3 \cdot 10^{18}$ neutrons/cm^2).

A technologically important property of DFO is the small weight loss
(small gas evolution) at high temperatures. Thus, for example, weight
losses in the air after 3000 hours at 210° were only 0.05%. Pressed DFO
specimens took up 0.5—0.8% moisture after 24 hours' boiling in water.

Ebonite specimens similarly treated showed a 1.4% increase in weight, while a capron specimen underwent decomposition. It would appear that the water is adsorbed on the surface only, since the original weight of the DFO specimens could be reestablished by merely drying at 80° for 15 minutes.

TABLE 38. Variation in the mechanical properties of a number of polymers under γ-irradiation

Polymer	Dose, Mrad	Variation in properties, % of initial value		
		tensile strength	elongation	elasticity modulus
Nylon /98/	17.6	−17	−3	−33
	10,000	−100	−100	−
Polyester /98/	30	−11	0	−7
	60	−7	+20	−14
	90	−32	+13	−14
Polythiazole /98/	60	−3	−32	+33
	90	−6	−84	+46
DFO polyimide /34/	1,000	0	0	0
	10,000	0	−20	0

At high temperatures (400−450°) DFO becomes cross-linked; at higher temperatures it gives a dense coke residue (coke number 0.6−0.65). The stability to chemicals and oils is the same as for all aromatic polyimides. The electrical properties of DFO are also typical for polyarimides: the tangent of dielectric loss angle between 20 and 200° is $1 \cdot 10^{-3}$ to $3 \cdot 10^{-3}$, the volume resistivity is about 10^{17} ohm·cm at 20° and $2 \cdot 10^{14}$ ohm·cm at 200°, and the electric strength of the films is 100−200 kV/mm.

The combination of the high thermostability, temperature resistance, mechanical and electrical properties, stability to radiation, etc., and the fact that it can be molded into bulk objects renders DFO a very promising material for practical utilization, particularly in high-temperature, radiation-resistant air-tight electrical installations.

It is seen that polyimides may be used to produce thermoplastic resins with a much higher temperature resistance and thermal stability than those at present available. Moreover, these properties of DFO do not represent the upper limit of the performance attainable by polyimides. In particular, the upper limit of working temperatures withstood by this material is not constituted by its thermostability, but by its temperature resistance. The temperature of prolonged constant quality performance of a thermoplastic polyimide with a softening point above 350° might be as high as 250−300°. Such representatives of thermoplastic polyimides can be discovered; this is indicated, for example, by the fact that it is possible to raise the temperature resistance by altering the chemical structure of the hetero-groups in the polyimide chain. Thus, if the −O− ether group in the dianhydride component is replaced by benzophenone −CO− and sulfone −SO₂−, the softening point increases to 290 and 340° respectively (polymers VI-5, IV-5, V-5 in Table 28), while the pressing capacity is retained.

Interesting results can be obtained with polyimides with aromatic links in the meta-position, For example, DFM polyimide (VI-3) yields extruded samples with a softening point of about 490°; samples pressed with glass fiber have a high impact strength (Table 37). DFM polymer is distinguished by its ability to crystallize. As has been noted, the cross-linking reactions in this polymer are much more intense than in other representatives of Group D. During pressing at 490—500° the crystallites melt, the polymer becomes cross-linked until high values of elasticity modulus are obtained and remains amorphous on being cooled, with a high softening point. Samples of plastics with a softening point of 430—450°, with a high tensile strength ($400-600$ kg/cm^2) at 400°, could also be prepared by pressing DFFG polyimide (III-9).

The difficulties involved in the processing of nonsoftening or weakly softening polyimides, in particular polypyromellitimides, are due to their intense cross-linking at the high temperatures at which they have to be processed, but the final product may well display the optimum possible performance in this group of compounds. Accordingly, this problem is being intensively studied. A promising technique is to process the poly-imides in an inert medium, since then the cross-linking proceeds at a slower rate (Figure 58).

Freeman et al. /64/ attempted to prepare a fiberglass plastic in which the binder is constituted by polypyromellitimide I-7 and polypyromellitami-doimide of the structure

The solution of polyamido-acid was used to impregnate a glass cloth which was then dried, heat-treated and pressed to obtain laminated fiber-glass plastic. The optimum mechanical parameters were displayed by products with the maximum content of the binder. In choosing the pressing temperature the authors were guided by the temperature variation of the tangent of dielectric loss angle tan δ of polymer films, on the assumption that a sudden increase of this parameter at high temperatures is an indication of the passage to a softened state.* The pressing temperature was 285° for the polyamidoimide and 365° for the polyimide. In order to facilitate the pressing, a combined binder was also tested; the glass cloth covered by the polyimide was then coated with polyamidoimide and pressed at the softening temperature of the latter. Samples of fiberglass plastics were tested for the stability of their mechanical and electrical parameters by prolonged heating in the air at 315 and 344°. The results of the testing at 315° are shown in Figure 100. It is seen that the variations in the mechanical properties were least for the fiberglass plastic containing pure polyimide. The polyamidoimide binder results in a higher initial tensile strength, but a much poorer thermostability. The combined binder also yields a less thermostable product than the pure polyimide. In all cases the tensile strength values of new fiberglass plastics at room temperature were much lower than those of fiberglass plastics based on

---

* It follows from /58/ that this is not always so. A sharp rise in tan δ, especially above 300°, may also be due to other causes, such as the appearance of conductance at high temperatures.

phenol resins, but unlike in the latter, they remained stable at high temperatures. The service life of a fiberglass plastic with polyimide binder at 315° is 1000 hours, and at 344° it is 300 hours, the service-ability criterion being a flexural strength value of 700 kg/cm². Fiber-glass plastics with a phenol binder had a service life of only 150 hours even at a much lower temperature (260°).

The low tensile strength values obtained /64/ for purely polyimide fiberglass plastic are very probably connected with the low degree of compression of the polyimide employed. This is also indicated by the intensive absorption of moisture and by the major changes in the electric parameters of the fiberglass plastic as a result of such absorption. These changes disappear, however, if the product is dried to its initial moisture content. The dielectric constant of the fiberglass plastic with polyimide binder is about 4.1, and the tangent of dielectric loss angle is $0.004 - 0.007$. These parameters (determined at a frequency of 10^{10} sec^{-1} remain practically unchanged between 20 and 315°. According to /64/ polyimide binder was the first to ensure a stable mechanical strength of the fiberglass plastic above 300° and the material has been recommended for use as radome of radio instruments in airplanes.

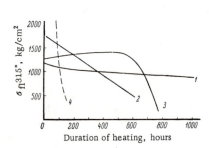

FIGURE 100. Variation of flexural strength at 315°C after prolonged heating in the air at 315°C for a number of fiberglass plastics /64, 65/.

Binders: 1 — polyimide; 2 — polyamido-imide; 3 — mixture of the two; 4 — phe-nol resin (tested and aged at 260°C).

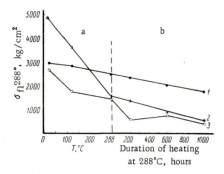

FIGURE 101. Flexural strength as a function of temperature (a) and of the time of holding at 288°C in the air (b) for a number of fiberglass plastics /107/.

Binders: 1 — polyimide; 2 — thermosetting epoxide resin; 3 — silicone resin.

The data reported by Todd et al. /107/ are also a good illustration of the advantages offered by fiberglass plastics with polyimide binder at high temperatures. If the softening point of the binder is around 400°, the strength of the fiberglass plastic varies much less with the temperature than does that of fiberglass plastics with silicone and epoxide binders. As a result, the strength of the former above 200° is 50% higher than that of the latter. The difference in the mechanical strengths becomes even larger after 1000 hours' heating in the air at 288° — the polyimide fiberglass plastic retains its flexural strength of 1700 kg/cm², while that of silicone and epoxide plastics is only 400 kg/cm² (Figure 101).

A comparison of the values of mechanical parameters of polyimide fiberglass plastics (/64, 107/ and Table 37) shows that the highest strength parameters will be obtained if the binder consists of a softening polyimide.

DuPont is now offering for sale /68, 73, 90, 107, etc./ laboratory quantities of a polyimide plastic product by the name of "Vespel" of two kinds: SP-1 (nonfilled) and SP-2 (filled with 15% graphite) as ready-made blocks to be converted into the desired object by mechanical working. The technological preparation and the chemical structure of this product have not been disclosed. Plastics SP-1 and SP-2 are intended for prolonged services at temperatures of up to 260° in the air and up to 315° in an inert atmosphere. They may also be utilized at temperatures up to 500° for brief periods of time. They retain a fair amount of flexibility at low temperatures (−200°C).

A synopsis of the physical and mechanical parameters of SP-1 is given in Table 39. The values of the elasticity modulus at different temperatures indicate that the polymer which constitutes SP-1 has a softening point below 400°. The product has a low moisture absorption, is insoluble in organic solvents, is resistent to acids, but is decomposed by strong alkalis and superheated steam. After 100 hours' boiling in water 60% of the initial strength is lost. The amounts of volatiles evolved in vacuo is small, and the product can withstand doses of fast electron irradiation of above 1000 Mrad without undergoing threshold changes. The plastic material SP-1 was used in experimental manufacture of instrument parts used in electrical and radio engineering for use in the outer space and in atomic reactors, piston gaskets in compressors, parts of fuel ducts of jet engines working at 232° and under a pressure of 200 atm, parts of systems containing liquid gases, diamond abrasive wheels, etc. Tests carried out on these objects confirmed that the material is highly promising.

A major problem in modern technology is the manufacture of solid antifrictional materials which do not require lubricants. This is necessary, first and foremost, in installations operating in vacuo and at high and low temperatures. The problem was solved by using polymeric materials filled with inorganic powders of the type of graphite, molybdenum disulfide, etc.

TABLE 39. Properties of unfilled polyimide plastic SP-1 /73,107/

Parameter	Numerical value
Density at 20°C, g/cm^3	1.42
Tensile strength, kg/cm^2, at:	
20°	875
150	620
250	500
315	320
400	220
Elongation-at-break at 20°C, %	6 − 8
Elasticity modulus, kg/cm^2, at:	
−190°	33,000
20	29,000
200	14,200
250	13,400
400	600
Shearing stress at 20°C, kg/cm^2	750
Compressive strength at 20°C, kg/cm^2	1550
Softening point under load of 200 kg/cm^2, °C	250
Linear expansion coefficient, degree^{-1}, between −180 and +260°C	$7 \cdot 10^{-5}$
Dielectric constant at 20°C and 10^6 hz	3.4
Volume resistivity at 20°C, ohm·cm	10^{17}
Tangent of dielectric loss angle at 20°C and 10^6 hz	0.003
Arc resistance, seconds	230 (phenol resins 5)

TABLE 40. Properties of polymeric anti-friction materials /41,107/

Parameter	Nylon	Teflon	Nonfilled SP-1 polyimide	Filled SP-2 polyimide
	filled			
Critical temperature, °C	200	260	320	320
Friction coefficient	0.1 — 0.2	0.04 — 0.25	0.05 — 0.2	0.04 — 0.2
Maximum pressure P, kg/cm^2	340	140	700	700
Maximum rate V, m/sec	10 — 20	50	50	50
Maximum PV index	1.4	11	36	36
PV index for prolonged service in the air (wear less than 0.1 μ/hour)	0.36	1.8	0.1	1
Wear, μ/hr, at: $PV = 9$	Do not work		7.6 (in the air) 1.3 · 10^{-2} (in nitrogen)	2.5 (in the air) 1.3 · 10^{-2} (in nitrogen)
$PV = 36$	" "		—	5.1 · 10^{-2} (in nitrogen)

TABLE 41. Friction and wear of polymeric materials under a high vacuum /50/.

Index	Teflon against stainless steel			Polyimide against stainless steel			Polyimide against polyimide
	nonfilled	25% glass (fiber)	25% copper (powder)	nonfilled	15% graphite (powder)	20% copper (fiber)	
Friction coefficient ...	0.28	0.3	0.28	0.18	0.25	0.05	0.5
Wear of abrading body, cm^3/hr	1.7 · 10^{-2}	1.7 · 10^{-4}	1.7 · 10^{-4}	8 · 10^{-4}	8 · 10^{-3}	1.3 · 10^{-4}	2 · 10^{-6}

N o t e. Test conditions: vacuum 10^{-9} mm Hg, abrader: a cylinder of 0.5 cm radius, substrate: rotating disk, rate of abrasion 2 meters/second, load 1 kg, no external heating of specimen.

/41/. The utilization of polyimides for the purpose considerably improves the service properties of autolubricating polymeric materials (Table 40); a considerable increase is noted in the values of maximum permissible working temperatures and maximum permissible loads and velocities. A noteworthy feature is the exceedingly small wear of polyimide materials in an inert atmosphere (dry nitrogen) under large loads and at high velocities, i.e., under conditions in which Capron and Teflon-based antifrictional materials, which are less heat-resistant and which display cold flow, cannot be employed at all. Since the volatility of the polymers in high vacuum is low, and their friction coefficients are small, they are particularly suitable for use as antifrictional materials in the outer space. However, antifrictional materials for use in the outer space must meet many other additional requirements such as the maximum possible mechanical strength, temperature resistance, thermal stability, etc. Buckley and Johnson /50/ studied the performance of various antifrictional polymeric materials, including block polyimide, under conditions of cosmic vacuum in the outer space ($10^{-7}-10^{-9}$ mm Hg). The magnitudes determined included the rate of loss in weight on heating, friction coefficient and wear. It was found that polyimide materials are much superior in these respects to materials based on Teflon. Specially interesting are data on the friction and wear (Table 41) from which it is seen that polyimide may be used to make materials with very low friction coefficients and a low wear under conditions of cosmic vacuum.

The optimum antifrictional characteristics were displayed by polyimide filled with copper fiber. The lowest wear was displayed by the pair polyimide - polyimide. Unfortunately, the authors /50/ give no data as regards friction in vacuum at elevated temperatures, when the advantages of polyimide materials may be expected to be especially striking.

Patents have been taken out for the manufacture of polyimide foam plastics, which are of high interest as thermal insulation and as lightweight construction materials with high maximum working temperatures.

Thus, practically all the main types of plastic materials may be prepared from polyimides.

Polyimide varnishes and adhesives

This field of application of polyimides was the first to be investigated. In 1960 — 1962 the results of tests performed on experimental samples of polyimide varnishes, which are now used in the manufacture of a set of electrical engineering materials known as "Pyre ML," were published by DuPont. The set includes impregnating and adhesive compositions, varnished glass fiber and polyimide enamel insulated wires. The tests performed on the materials themselves and on experimental electric machines were the first indication of the potential value of polyimides in industry.

Typical parameters of "pyre ML" varnish compositions are shown in Table 42.

The varnishes may be stored /97/ in moisture -proof vessels for more than one year at 0 — 4° and for 3 — 5 months at 20°. They are applied by rollers, by spraying or dipping. Impregnation of complex objects should be carried out in vacuo; depending on the coating technique, the concentration of the varnish solution should be varied by the addition of solvent. The duration of the first baking stage — removal of solvent — which is carried out at 120 — 180°, should also be different for different kinds of objects. It is pointed out /107/ that the varnishes showed satisfactory adhesion to almost all metals: copper, brass, cast iron, chemically treated aluminum and steel, etc. The quality of cohesion between metals produced by polyimide varnishes may be illustrated by the strength of cohesion of stainless steel (Figure 102). The bonding was performed under a pressure of $14 \, kg/cm^2$ for 2 hours at 300°. The polyimide bonding strength at room temperature was lower than that displayed by an epoxy-phenol resin (150 against $300 \, kg/cm^2$), but when the temperature was increased and especially at long holding times under these conditions, the variation of the polyimide bonding strength was much less. The epoxy-phenol bond is disrupted and does not withstand applied mechanical loads for several hours at 290 and 370°, while the bonding strength of the polyimide remains unchanged after hundreds of hours of aging at these temperatures.

In many practical applications, such as the utilization of polyimide varnishes to impregnate windings of electric machines, it is very important for the varnish not only to have a high bonding power but also not to soften at elevated temperatures. This is particularly important in DC collector machines which often break down due to the adherence of coal dust to the winding /75/.

Enamel-insulated coil wires are the most modern kind of conductor materials for small and medium -powered electric machines and instruments. The most important quality requirements which the enamel insulation has to satisfy under the service conditions of these installations

135

TABLE 42. Properties of "Pyre LM" varnish, dry residue and coatings/ 97, 107 /

Type of varnish	Dry residue, %	Viscosity at 20°C, poises	Solvent	Weight loss by dry residue, %			Baked in thin layer	Glass transition temperature, °C	Tensile strength, kg/cm²	Elongation-at-break, %	Dielectric constant at 20°, 10³ hz	Dielectric losses at 10³ hz	Electric strength, kV/mm
				at 300°C during 50 hours	at 400°C during 200 minutes	at 450°C during 200 minutes							
A	16.5	70	MP/DMA	1	11	20	30 min, 150°;	500	1050	20	3.5	0.002	175
B	25	35	DMF	1	12	35	2 min, 410°.	300	900	11	2.0	0.002	160
C	25	8	DMF	4	11	46	1 hr, 120°;	400	700	7	2.7	0.001	130
D	60	80	MP/KS	5	12	31	1 hr, 150°; 1 hr, 230°.	—	500	3	2.8	0.001	175

Note. MP — N-methylpyrolidone; DMA — dimethylacetamide; DMF — dimethylformamide; KS — xylene.

concern tensile strength, elasticity, resistance to wear and to perforation by concentrated loads, electric resistance and electric strength. The quality specifications concerning the temperature resistance and thermal stability of enamel insulations are now particularly stringent; this is due to the fact that the articles must give good service at high temperatures and also because it is desired to improve their resistance to overloading and their specific output. The scope of application of organic enamel insulation is thus greatly extended by the use of polyimides.

Polyimide-enameled wires may be produced on conventional installations /75, 97/. Enamel coatings may be applied to copper, aluminum, nickelized copper and steel wires, to wires covered with glass and ceramics etc. The wire cross-section may be round or rectangular. The thickness of the insulating layer may vary between 2 and $20\,\mu$. The coating is applied by passing the wire through a varnish bath. The excess varnish is removed with callipers or felt shingles. In the latter case the varnish concentration must be lower and the number of passes must be increased. Drying and curing of the coatings is performed in vertical and horizontal furnaces. The velocity and temperature conditions are so selected that the already coated and cured wire might be wetted by the varnish on passing through the bath without destroying the coating formerly applied. The final curing of the wire may be performed in these furnaces or else by heating on metal coils. The production of polyimide wires is straightforward except that toxic and corrosive solvents must not be employed.

As regards their working performance, polyimide-enamelled wires are far superior to any others now known /7, 68, 73, 90, 107, etc./. An exception is the frictional strength (test with charged needle sliding along wire surface) of polyimide coatings which is slightly inferior to that of the well known polyester and Viniflex, but here, too, polyimide-coated wires will give high quality performance, provided the conditions of their thermal treatment are suitable chosen.

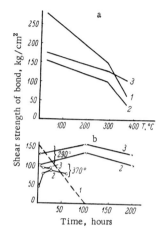

FIGURE 102. Bonding strength as a function of temperature (a) and time of dwelling in the air at 290 and 370°C (b) / 107/.

Adhesives: 1 — epoxy-phenol resin; 2 — pure polyimide; 3 — filled polyimide.

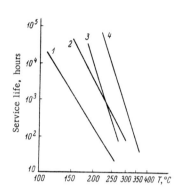

FIGURE 103. Service life of enamel wires. Testing of coils under different thermal stresses / 24, 75, 97/.

Insulation: 1 — Formuar; 2 — polyester; 3 — organosilicon; 4 — polyimide.

Polyimide insulation markedly prolongs the service life of electric wires at high temperatures (Figure 103). At 250° the service life is 10,000 hours for polyimide wires as against 200 — 600 hours for wires coated with organosilicon varnishes. The resistance of polyimide enamels to perforation is considerable. Thus, it has been shown /18/ that enamel insulation made of PM polypyromellitimide was not perforated at 310° even when the metal of the wire was fully disintegrated, whereas polyester insulation became perforated under one-fourth of the pressure even at 200 — 220°. Polyimide insulation is very highly resistant to thermal impact; it does not crack at the bends when rapidly heated to 500° or more. Teflon-coated wire withstood this test only up to 325°, while polyester-varnished wire only up to 240°. Polyimide insulation is elastic, does not crack when sharply bent or at below-zero temperatures, down to —195° and even lower. When heated to 300°, resistivity and the breakdown voltage do not show marked changes. It has good continuity and is perfectly resistant to varnishes.

Vasil'eva and Shakai /7/ published interesting results of standard tests on experimental heat-resistant enameled wires of Soviet manufacture. Wires, 0.4 mm in diameter, of the following brands were tested: PNET (insulation on polysiloxane varnish), PEKF (Ftoroplast insulation on ceramic support; and PM (insulation made of polypyromellitimide). The purpose of the tests was to evaluate the performance of these wires when

137

used as windings in electrical instruments under service conditions involving high standards of elasticity and strength under concentrated loads at elevated temperatures. The elasticity was taken to be the smallest curvature radius of the wire withstood by the insulation after high-temperature aging ("thermal stability" test) and the smallest curvature radius of the bent wire at which a rapid heatup of the bent portion of the wire to a high temperature is not accompanied by the decomposition of the insulation ("thermal impact" test). Resistance to perforation was determined as the temperature at which electric contact was established between the core of the wire and an indentor of 0.5 mm radius pressing on it with a force of 0.65 kg. The results of the test (Table 43) indicate that polyimide-insulated wires are much superior as regards their elasticity and resistance to perforation.

TABLE 43. Results of tests carried out on enameled wires /7/

Brand of wire	"Thermal impact" — minimum value of D/d after rapid heatup and holding			"Thermal stability" — minimum value of D/d after holding			"Plasticity" — perforation temperature, °C
	250°, 1 hr	300°, 1 hr	400°, 1 hr	24 hr, 250°	1000 hr, 250°	24 hr, 300°	
PNET	7	7	12	12	—	25	120
PEKF	2	2	—	2	5	2	220
PM	1	1	1	2	2	2	300 — 400

Note. D — wire diameter; d — diameter of rod around which the wire was bent.

The continuity of polyimide insulation on a wire bent around a rod of a diameter equal to that of the wire, remained unimpaired on thermal impact up to 400°. The insulation retained its capacity to withstand sharp flexure after prolonged aging at 250 and 300°. As regards the perforation temperature, it could only be established that it was much above 300°, since the core of the wire disintegrated prior to any perforation of the insulation. Polysiloxane insulation (PNET wire) proved to be much less elastic in the intial state and to lose its elastic properties rapidly on aging and at elevated temperatures of thermal impact (the minimum flexure radii of the wire were one order higher). Plastic perforation of the insulation took place even at 120°. Intermediate results were obtained for wire with combined enamel insulation (PEKF wire), but this was the result not so much of the presence of the enamel coating as of the presence of the ceramic support.

In view of the above, the authors /7/ laid special stress on the desirability of using polyimide enamels for wires with rectangular cross-section. In this case the enamel insulation must satisfy particularly stringent specifications: in making rectangular wire windings of large electrical machines, the wire must be bent "edge-on" (the insulation must be flexible); also, major contact stresses are produced between contiguous parts of the coil by centrifugal forces during exploitation (insulation must not become perforated).

The properties of "Pyre ML" polyimide varnished glass have been described /75/. It is prepared by impregnating glass cloth with poly-amido-acid varnishes with subsequent curing. It is mainly used as lining, e. g., groove electric insulation. Pure polyimide film may be employed for the same purpose.

The polyimide materials used in electrical engineering which have just been described were effectively tested on model windings of electrical machines and on operative electric motors /104/. Thus, for example, the following experiments were performed. The electric motor was tested heavily overloaded under throttling conditions. After 30 minutes the temperature of the windings attained 600°, the collector melted and the oil caught fire. At the end of the test it was found that the resistance and the electric strength of the polyimide insulation remained unchanged. In another experiment such electric motors were run for three months at 240 and 260° without any damage to the insulation; heat-resistant electric motors at present on the market with enameled wire coils can work under these conditions for not longer than 100—200 hours. These experiments immediately show that the use of polyimide materials as insulation in electrical machines makes it possible to work them at much higher loads and temperatures, increase their specific power (1.5—2 times, according to /107/), ensure working performance under specially difficult conditions (e.g., at a high temperature and under irradiation simultaneously), prolong their service life and limit the consumption of ferrous and nonferrous metals to a considerable extent. Obviously, these considerations also apply to any other electrical engineering installations, in which the utilization of polyimides is possible and expedient.

Polyimide fibers

Increasing the temperature resistance and thermal stability of technological fibers is one of the main problems in this field of polymeric materials. The latest major advance in this respect was the development of aromatic polyamide fibers /81/ which have a high strength up to 300°, and which retain their mechanical properties for prolonged periods of time at 200°. The use of polyimides results in fibers which have even better parameters. In its intermediate (polyamido-acid) form this type of polymer can be readily processed into fibers by the conventional "dry" or "wet" methods. As has been seen above, hot-drawn polyimides have superior mechanical strength parameters and elasticity moduli. Oriented polyimide films, heat-treated at high temperatures, do not contract during heating. This means that there is a real possibility of producing high-modulus, nonshrinking polyimide fibers. Research work on this question outside the Soviet Union is mainly conducted by DuPont /55/, who have taken out several patents for different methods of manufacture of polyimide fibers.

Rudakov et al. /32/ prepared and studied the properties of laboratory specimens of fibers made of polypyromellitimide PM. The fibers were prepared by the dry method. A 20—30% solution of the polyamido-acid (viscosity ~ 500 poises) was extruded through a single needle-shaped die. The resulting fiber was passed through a drying chamber and wound on a metal bobbin. The linear rate of rotation of the bobbin was so adjusted that the fiber was stretched by some 20,000%, the resulting fiber diameter being about 20μ. The fiber on the bobbins was heated to 250°, hot-stretched in helium to 150—180% at 300—350°, and was then cured without stretching at 400°.

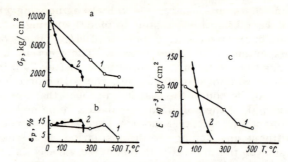

FIGURE 104. Tensile strength (a), elongation-at-break (b) and elasticity modulus (c) as functions of the temperature for single fibers of polyimide PM (1) and poly(vinyl alcohol) cord of "E" brand (2) /32/ .

PM fiber which had not been hot-stretched, but had been cured at 400° on bobbins, was strong and elastic; tensile strength 2500 kg/cm², elongation-at-break 50—80%, "knot" strength 80—100% of the initial.

PM fiber after hot stretching was much stronger and had a much higher elasticity modulus, but was less elastic. In its mechanical properties at room temperature it resembled industrial cord fibers, but retained its working performance at much higher temperatures (Figure 104). For example, at 350—400° it had the same values of mechanical parameters as the high quality "E" cord, made of poly(vinyl alcohol), at 200°.

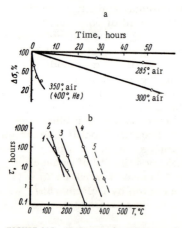

FIGURE 105. Variation of tensile strength of polyimide PM fibers with time at various temperatures in the air and in helium (a) and thermostability index τ (time prior to a 50% decrease in strength) as a function of the temperature (b) for a number of fibers /32/

1 — poly(vinyl alcohol) fiber; 2 — Capron; 3 — Terylene; 4 — polyimide PM, weight in the air; 5 — polyimide PM, in helium.

The mechanical parameters of polyimide fibers are much more thermostable than those of Capron and polyester fibers (Figure 105). The stable performance temperature — defined as the duration of aging τ which results in a 50% decrease in tensile strength — is at least 100—150° higher. It is also seen from the figure that polyimide fiber has a long service life at 250° in the air.

An important property (manufacture of tire cords, transmission belts, etc.) of polyimide fibers is that they do not shrink on being heated. The result is that the stresses produced when the fixed polyimide fibers are heated are hundreds of times smaller than those in strongly oriented fibers made of typical linear polymers which soften or melt at high temperatures (Figure 106). This means that objects reinforced with these fibers will keep their shape at elevated temperatures.

Fibers with very interesting properties were prepared from the aromatic polyimide PFG /20/ by Frenkel' et al. Their mechanical properties (Table 44) are

140

comparable only with those of glass and metal threads. Especially
striking are the high values of elasticity modulus of these fibers, which
attain 380,000 kg/cm^2 at room temperature; this may be compared with
500,000 — 700,000 kg/cm^2 for glass, 400,000 kg/cm^2 for magnesium,
700,000 kg/cm^2 for aluminum and 1,000,000 kg/cm^2 for copper.

TABLE 44. Physical and mechanical properties of PFG polyimide fiber

Temperature, °C	Strength, kg/cm^2	Elongation, %	Elasticity modulus, kg/cm^2
−195	18,000	5 − 6	−
+ 20	11,700	4 − 5	380,000
+ 200	6,600	2 − 2.5	330,000
+ 300	5,000	1.5 − 2	300,000
+ 400	6,000	4 − 5	240,000

Polyimide fibers with a high elasticity modulus can be expected to
furnish a superior material for the manufacture of high-temperature cord
and especially filled plastics. These
materials, which are made of polyimide fiber and polyimide binder, may
be expected to display not only thermostability, but also low density and uni-
formity (radio-transparency), which
are very important parameters in many
applications. The elastic polyimide
fibers may be used to manufacture
fabrics which retain their flexibility
both at very low and very high temperatures, are resistant to UV radiation
and other cosmic effects. They may
be used as electrical and thermal insulation etc. This is indicated both by the
data which have just been quoted for
polyimide fibers and by the properties of polyimides in general.

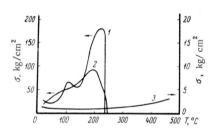

FIGURE 106. Stresses resulting on heating rigidly fixed fibers made of poly(vinyl alcohol)
(1), Capron (2) and PM polyimide (3) fibers
/ 32/ .

Modern application of polymeric materials during the past 10−15
years require that their temperature resistance be much increased.
This has affected the course of development of a large branch of high-
polymer chemistry. The production of ring-chain polymers, which can
be processed and which display an ensemble of valuable physical and
mechanical properties, represents an important advance in this respect.
Polyimide polymers described in this book, as well as many other ring-
chain polymers, are certain to be widely utilized in the nearest future
in the manufacture of various materials which ensure the advance of modern
technology. They are also of high purely scientific interest, and pure
research on the subject is of importance in further development of physics
and chemistry of high-molecular compounds.

BIBLIOGRAPHY

PUBLICATIONS IN RUSSIAN

1. ADROVA, N.A., M.M.KOTON, and E.M.MOSKVINA. — Doklady AN SSSR, 165 (5):1069. 1965.
2. ARDASHNIKOV, A.Ya., E.I.KARDASH, B.V.KOTOV, and A.N.PRAVEDNIKOV.— Doklady AN SSSR, 164 (6):1293. 1965.
3. BELLAMY, L.J. The Infrared Spectra of Complex Molecules. [Russian translation, 1963.]
4. BESSONOV, M.I. — Vysokomolekulyarnye Soedineniya, 8 (1):66. 1967.
5. BERLIN, A.A., B.I.LIOGON'KII, G.M.SHAMRAEV, and G.V.BELOV. — Izvestiya AN SSSR, seriya khimicheskaya, No.5:945. 1966.
6. BOLDYREV, A.G., N.A.ADROVA, M.I.BESSONOV, M.M.KOTON, E.V. KUVSHINSKII, A.P.RUDAKOV, and F.S.FLORINSKII. — Doklady AN SSSR, 163 (5):1143. 1965.
7. VASIL'EVA, A.P. and S.F.SHAKAI. — Aviatsionnaya Promyshlennost', No.8. 81. 1965.
8. VINOGRADOVA, S.V., V.V.KORSHAK, and Ya.S.VYGODSKII. — Vysokomolekulyarnye Soedineniya, 8 (5):809. 1966.
9. GUSEV, M.P. and E.ABDULA-ZADE. — Khimiya i Tekhnologiya Polimerov, No.11:93. 1965.
10. DZHOUNS, I., F.OCHINSKII, and F.RAKLEI. — Khimiya i Tekhnologiya Polimerov, No.8:30. 1963.
11. ZHURKOV, S.N. and G.Ya.RYSKIN. — ZhTF, 24 (5):797. 1954.
12. ZAKOSHCHIKOV, S.A., K.N.VLASOVA, G.M.ZUBAREVA, N.M.KRASNOVA, and G.A.RUZHENTSOVA. — Plasticheskie Massy, No.1:14. 1966.
13. ZAKOSHCHIKOV, S.A., G.M.ZUBAREVA, and G.M.ZOLOTAREVA. — Plasticheskie Massy, No.4:9. 1966.
14. ZAKOSHCHIKOV, S.A. and G.A.RUZHENTSOVA. — Vysokomolekulyarnye Soedineniya, 8 (7):2131. 1966.
15. IOFFE, B.V. Refraktometricheskie metody khimii (Refractometric Methods in Chemistry). — Moskva, Goskhimizdat. 1956.
16. KOLESNIKOV, G.S., O.Ya.FEDOTOVA, E.I.KHOFBAUER, and KHUSEIN-KHAMID-MOKHAMMED ALI AL'-SUFI. — Vysokomolekulyarnye Soedineniya, 8 (8):1440. 1966.

17. KORSHAK, V. V. and S. V. VINOGRADOVA. Poliarilaty (Polyarylates), p.53. — Moskva, Izd."Nauka." 1964.

18. KOTON, M. M., A. P. RUDAKOV, M. I. BESSONOV, L. A. LAIUS, F. S. FLORINSKII, and S. F. SHAKAI. — Aviatsionnaya Promyshlennost' No. 1:26. 1965.

19. KOTON, M. M., B. I. YAKOVLEV, A. P. RUDAKOV, T. S. KNYAZEVA, F. S. FLORINSKII, M. I. BESSONOV, M. M. KULEVA, G. A. TOLPAROVA, and L. A. LAIUS. — ZhPKh, 38 (12):2728. 1965.

20. KOTON, M. M., S. Ya. FRENKEL', and A. P. RUDAKOV. — Vestnik AN SSSR, No. 8:56. 1966.

21. KUDRYAVTSEV, V. V., M. M. KOTON, E. I. POKROVSKII, and E. F. FEDOROVA. — Tezisy XIII Nauchnoi Konferentsii IVS AN SSSR, p.4. Leningrad, Izd. "Nauka." 1965.

22. KEY, D. and T. LEBY. Tables of Physical and Chemical Constants. [Russian translation, 1962.]

23. LAIUS, L. A., F. S. FLORINSKII, M. I. BESSONOV, E. V. KALLISTOVA, and N. A. ADROVA. — Tezisy XIV Nauchnoi Konferentsii IVS AN SSSR, p.25. Leningrad, Izd. "Nauka." 1967.

24. NISHIZAKI, SUMI. — Denki Keisan. 33 (5):671. 1965.

25. PIMENTELL, G. C. and A. L. McCLELLAN. Hydrogen Bond. [Russian translation. 1964.]

26. PLONKA, Z. Yu. and V. M. AL'BREKHT. — Vysokomolekulyarnye Soedineniya, 7(12):2177. 1965.

27. RAZVODOVSKII, E. F. — Khimiya i Tekhnologiya Polimerov, No. 12:62. 1964.

28. RAZVODOVSKII, E. F. — Khimiya i Tekhnologiya Polimerov, No. 3:63. 1965.

29. RENNE, V. T. Plenochnye kondensatory s organicheskim sinteticheskim di-elektrikom (Film Condensers with Synthetic Organic Dielectric), p.174. Moskva, Gosenergoizdat. 1963.

30. RUDAKOV, A. P., M. I. BESSONOV, N. A. ADROVA, and M. M. KOTON. — Doklady AN SSSR, 172 (4):899. 1967.

31. RUDAKOV, A. P., M. I. BESSONOV, M. M. KOTON, E. I. POKROVSKII, and E. F. FEDOROVA. — Doklady AN SSSR, 161 (3):617. 1965.

32. RUDAKOV, A. P., M. I. BESSONOV, M. M. KOTON, and F. S. FLORINSKII. — Khimicheskie Volokna, No. 5:20. 1966.

33. RUDAKOV, A. P., M. I. BESSONOV, M. M. KOTON, and F. S. FLORINSKII. — Elektrichestvo, No. 11:86. 1966.

34. RUDAKOV, A. P., F. S. FLORINSKII, M. I. BESSONOV, K. N. VLASOVA, M. M. KOTON, and P. M. TANUNINA. — Plasticheskie Massy, No. 9:26. 1967.

35. SMIRNOVA, V. E. — Tezisy XIV Nauchnoi Konferentsii IVS AN SSSR, p.3. Leningrad, Izd. "Nauka." 1967.

36. SIDOROVICH, A. V., M. I. BESSONOV, M. M. KOTON, and A. P. RUDAKOV. — Doklady AN SSSR, 165 (4):848. 1965.

37. TRELOAR, L. R. Physics of Rubber Elasticity. [Russian translation, 1953.]

38. AMBORSKI, L.E. — Ind. Eng. Chem., Prod. Res. Dev., 2 (3):189. 1963.
39. BELL, V.L. and G.F.PEZDIRTZ. — J. Polym. Sci., B3 (12):977. 1965.
40. BOGERT, T.M. and R.R.RENSHAW. — J.Am. Chem. Soc., Vol.30:114. 1908.
41. BOWER, G.R.— Metal Progress. 88 (3):106. 1965.
42. BOWER, G.M. and L.W.FROST. — J. Polym. Sci., A1 (10):3135. 1963.
43. British Patent No.941158. 1963.
44. BROWN, D.W. and L.A.WALL. — J. Phys. Chem., 62 (7):848. 1958.
45. BRUCK, S.D. — Am. Chem. Soc., Polym. Preprints, 6 (1):28. 1965.
46. BRUCK, S.D. — In: Vacuum Microbalance Techniques, ed. by P.M.Waters, Vol.4:247. N.Y. 1965.
47. BRUCK, S.D. — Polymer, 5 (9):435. 1964.
48. BRUCK, S.D. — Polymer, 6 (1):49. 1965.
49. BRUCK, S.D. — Polymer, 6 (5):319. 1965.
50. BUCKLEY, D.H. and R.L.JOHNSON. SPE Trans., 4 (4):306. 1964.
51. CAMPBELL, R. and A.CHENEY. — Mod. Plastics Encyclopedia, J. Whiley, p.244 N.Y. 1966.
52. Chem. Engn., 71 (11):98 (May 25). 1964.
53. Chem. Engn., 72 (15):100 (July 19). 1965.
54. Chem. Engn. News,42 (40):52 (Oct.5). 1964.
55. Chem. Engn. News, 42 (34):24 (Aug.24). 1964.
56. Chem. Week, 95 (14):71. 1964.
57. COLSON, I.G., R.H.MICHEL, and R.M.PAUFLER. — J. Polym. Sci., A-1, 4 (1):57. 1966.
58. COOPER, S.L., A.D.MAIL, and A.V.TOBOLSKY. — Text. Res. Journ., 35 (12):1110. 1965.
59. DAVANS, F. and C.MARVEL. — J. Polym. Sci., A3 (10):3549. 1965.
60. EDWARDS, W.M. and A.L.ENDREY. — Brit. Patent No.903271. 1962.
61. EDWARDS, W.M. and I.M.ROBINSON. — U.S. Patent No.2760835. 1955.
62. Electron Packag. a. Prod., 5 (4):74. 1965.
63. FLASHMAN, B.F. and P.S.SHENIAN. — Rubb. Plast. Age, 47 (4):383. 1966.
64. FREEMAN, I., L.FROST, G.BOWER, and E.TRAYNOR. — SPE Trans., 5 (2):75. 1965.
65. FREEMAN, I., E.TRAYNOR, I.MIGLARESE, and R.LUNN. — SPE Trans., 2 (3):216. 1962.
66. FROST, L.W. and I.I.KESSE. — J. Appl. Polym. Sci., 8 (3):1039. 1964.
67. GRESHAW, W.F. and M.A.NAYLOR. — U.S. Patent No.2731447. 1956.
68. Gummi-Asbest-Kunststoffe. 19 (2):143. 1966.
69. HEACOCK, I.F. and C.E.BERR. — SPE Trans., 5 (2):105. 1965.
70. HERMANS, P.H. and I.W.STREET. — Makromolec. Chem., Vol.74:133. 1964.
71. IKEDA, R. — J. Polym. Sci., B4 (5):353. 1966.
72. IMAI, I., K.UNO, and I.IWAKURA. — Makromolec. Chem., Vol.94:114. 1966.
73. Iron Age, 191 (20):88. 1963.

74. LADY, L.H., I.F.KESSE, and R.E.ADAMS. — J. Appl. Polym. Sci., 3 (7):71. 1960.

75. LEPPLA, R. and R.CARRYER. — Insulation, 9 (7):35. 1963.

76. LEUCHEUX, M.H. and A.D.BONDERT. — J. Polym. Sci., 48 (1):405. 1960.

77. LONCRINI, D.F. — J. Polym. Sci., A-1, 4 (6):1531. 1966.

78. LONCRINI, D.F., W.L.WALTON, and B.B.HUGNES. — J. Polym. Sci., A-1, 4 (2):440. 1966.

79. MADORSKY, S.L. — J. Res. NBS, 62 (6):219. 1959.

80. MADORSKY, S.L. and S.STRAUS. — J. Res. NBS, 63A (6):261. 1959.

81. McCUNE, L.K. — Text. Res. Journ. 32 (9):762. 1962.

82. MELVIN, B.I. and D.I.PARISH. — Insulation, 11(9):120. 1965.

83. Modern Plastics, 42 (5):104. 1965.

84. Modern Plastics, 43 (1):123. 1966.

85. NISHIZAKI, S. — J. Chem. Soc. Jap., Ind. Chem. Sec., 69 (7):1393. 1966.

86. NISHIZAKI, S. and A.FUKAMI. — J. Chem. Soc. Jap., Ind. Chem. Sec., 66 (3):382. 1963.

87. NISHIZAKI, S. and A.FUKAMI. — J. Chem. Soc. Jap., Ind. Chem. Sec., 67 (3):474. 1964.

88. NISHIZAKI, S. and A.FUKAMI. — J. Chem. Soc. Jap., Ind. Chem. Sec., 68 (2):383. 1965.

89. NISHIZAKI, S. and A.FUKAMI. — J. Chem. Soc. Jap., Ind. Chem. Sec., 68 (3):574. 1965.

90. Plastics Technology, 8 (12):26. 1962.

91. PLENIKOWSKI, S. and O.UMMINGER. — Kunststoffe, 55 (11):822. 1965.

92. POBINER, H. and M.JAWITZ. — J. Appl. Polym. Sci., 9 (3):1193. 1965.

93. PRESTON, I. and W.BLACK. — J. Polym. Sci., B3 (5):845. 1965.

94. PRINCE, M. and I.HORNYAK. — J. Polym. Sci., B4 (9):601. 1966.

95. REDDISH, W. — Trans. Farad. Soc., 46 (2):459. 1950.

96. SCALA, L. and W.HICKAM. — J. Appl. Polym. Sci., 9 (1):245. 1965.

97. SCHMIDT, K., G.NAUBERT, and H.ROMBRECHT. — Electrotechn. Zt., B15 (21):603. 1963.

98. SHEEHAN, W. — Polym. Eng. Sci., 5 (3):263. 1965.

99. SROOG, C.E. — Int. Sympos. Makromol. Chem., p.155. Praga. 1965.

100. SROOG, C.E., S.V.ABRAMO, C.E.BERR, W.M.EDWARDS, A.L.ENDREY, and K.L.OLIVER. — J. Am. Chem. Soc., Polymer Preprints, 5 (1):132. 1964.

101. SROOG, C.E., A.L.ENDREY, C.V.ABRAMO, C.E.BERR, W.M.EDWARDS, and K.L.OLIVER. — J. Polym. Sci., A3 (4):1373. 1965.

102. STRAUS, S. and L.WALL. — J. Res. NBS, 60 (1):39. 1958.

103. STRAUS, S. and L.WALL. — J. Res. NBS, 63A (3):269. 1959.

104. TATUM, W.E., L.E.AMBORSKI, C.W.GEROW, I.F.HEACOCK, and R.S.MALLOUK. — Electr. Insul. Conf. Mater.and Appl., 5-th Conf., Chicago, Ill., No.1:1. 1963.

105. TERUNOBI, U. — J.Polym. Sci., B3 (8):679. 1965.

106. TOBUSHI, I., K.TAGAKI, and R.ODA. — J. Chem. Soc. Jap., Ind. Chem. Sec., 67 (5):1081. 1964.

107. TODD, N., F. WOLFF, R. MALLOUK, and F. SCHWETZZER. — Machine Design., 36 (10): 228 (Apr. 23). 1964.

108. TOMAS, C.R. — Brit. Plastics, 38 (1): 36. 1965.

109. UNISHI, F. and T. TUYIMURA: — J. Chem. Soc., Jap., Ind. Eng. Sec., 68 (11): 2275. 1965.

110. VANDEUSEN D.L. — J. Polym. Sci., B4 (3): 211. 1966.

111. VOLLMERT, B. — Kunststoffe, 56 (10): 692. 1966.

112. WALLACH, M.L. — Am. Chem. Soc., Polym. Preprints, No. 1: 53. 1965.

113. WRASIDLO, W., P. HERGENROTHER, and H. LEVINE. — Am. Chem. Soc., Polym. Preprints, 5 (1): 141. 1964.

114. YOUNG, C.W. — J. Am. Chem. Soc., 69 (3): 1410. 1947.

Appendix I

PATENTS AND CERTIFICATES OF AUTHORSHIP ON THE
SUBJECT OF POLYIMIDES*

BRITISH PATENTS

1. No. 762,152 (1962). Preparation of polypyromellitimides.
2. No. 898,651 (1962). Preparation of polyamido-acids from various tetra-
    carboxylic acid dianhydrides and diamines.
3. No. 903,271 (1962). Preparation of polyimides by chemical treatment of
    polyamido-acids.
4. No. 903,272 (1962). Preparation of polyimides for the manufacture of films and
    coatings.
5. No. 904,599 (1962). Preparation of polyamido-imide resins.
6. No. 935,388 (1963). Preparation of polyimides with amido groups in the chain.
7. No. 941,158 (1963). Preparation of pressed aromatic polyimides with hetero-
    atoms and hetero-groups in diamine and dianhydride components.
8. No. 942,025 (1964). Preparation of polyimides based on benzophenone-tetra-
    carboxylic acid.
9. No. 945,673 (1964). Preparation of metal-containing polyimide films.
10. No. 945,674 (1964). Preparation of compositions containing polyamido-acids
    and their metal salts.
11. No. 972,007 (1964). Preparation of electroconducting polyimide materials
    and objects made of such materials.
12. No. 973,377 (1961). Preparation of ester-imide resin.
13. No. 980,273 (1965). Preparation of polyimides of different structures.
14. No. 980,274 (1965). Preparation of polyimides from polyamido-acids formed
    on mixing solutions of dianhydrides and diamines.
15. No. 980,855 (1965). Preparation of polyamido-acid powder.

* Most patents, both in the U.S. and in other countries, have been taken out by
  DuPont. The list is based mainly on the data published in Chemical Abstracts.

16. No. 982,914 (1965). Preparation of aromatic polyimides containing silicon, phosphorus and other elements in the diamine component.
17. No. 985,237 (1965). Preparation of polyimides of the type of poly(N-methyl-phthalimide).
18. No. 986,358 (1963). Preparation of polyimides by heating polyamido-acids in the presence of tertiary amine.
19. No. 997,391 (1962). Coating objects made of polyfluorocarbon compounds with polyimide varnish.
20. No. 999,578 (1965). Preparation of polyimide foam plastics.
21. No. 999,579 (1965). Preparation of polyimide foam plastics.
22. No. 1,009,211 (1965). Preparation of polyimide insulation varnishes.

BELGIAN PATENTS

23. No. 615,937 (1962). Preparation of two-layered enamel wire insulation with one polyimide layer.
24. No. 627,623 (1963). Preparation of polyimides by chemical treatment of poly-amido-acids.
25. No. 627,625 (1963). Preparation of polyamido-acid powders and their thermal and/or chemical treatment to polyimide.
26. No. 627,626 (1963). Preparation of pressed polyamide powders.
27. No. 630,749 (1963). Preparation of current-conducting polyimide films.
28. No. 638,688 (1964). Preparation of polyimide foam films.
29. No. 638,689 (1964). Preparation of polyimide foam plastics.
30. No. 641,568 (1964). Preparation of filled polyimide compositions for anti-frictional materials.
31. No. 649,336 (1964). Preparation of aromatic polyimides with fluorine and chlorine in the diamine component.

DUTCH PATENTS

32. No. 6,400,422 (1964). Preparation of polyamido-imides based on acyl halogeno-derivatives of trimellitic acid dianhydride and diamines.
33. No. 6,406,900 (1964). Preparation of polyimides based on fluorine-containing diamines.
34. No. 6,410,211 (1965). Preparation of polyimides from cyclopentanetetracarboxylic acid dianhydride and various diamines without utilization of solvents.
35. No. 6,412,661 (1965). Preparation of polyimides by thermal and chemical treatment of polyamido-acids.
36. No. 6,412,662 (1965). Preparation of polyimide-based adhesive compositions.
37. No. 6,413,550 (1965). Preparation of aromatic polyamido-esters and their conversion to polyimides by heating.
38. No. 6,413,551 (1965). Preparation of aromatic polycarboxyamides which can be converted to polyimides by thermal and chemical treatments.

39. No. 6,413,552 (1965). Preparation of polyamides which can be converted to polyimides by heating.
40. No. 6,414,419 (1965). Preparation of polyimides based on silicon-containing diamines.
41. No. 6,414,424 (1965). Preparation of polyimide copolymers by polycondensation of two dianhydrides and one diamine.
42. No. 6,414,673 (1965). Preparation of polyimides for wire coatings from carbonyl-diphthalic anhydrides.
43. No. 6,500,135 (1965). Preparation of polyimide intermetal bonds by curing mixtures of aromatic amido-esters.
44. No. 6,500,641 (1965). Preparation of polyimide coatings based on reactive multi-component compositions.
45. No. 6,504,004 (1965). Preparation of polypyromellitimide fibers by dry formation of polyamido-acid solutions and thermal treatment.
46. No. 6,506,280 (1965). Preparation of anti-frictional materials from polyimide and fluorinated resins.

SPANISH PATENTS

47. No. 288,912 (1963). Preparation of electric insulation phthalimide compositions.
48. No. 304,411 (1965). Preparation of high molecular weight polyamido-acids.
49. No. 304,412 (1965). Preparation of polyimides by thermal and chemical treatment of polyamido-acids.

U. S. PATENTS

50. No. 2,710,853 (1955). Preparation of polyimides by curing polypyromellitamido-acids.
51. No. 2,900,369 (1959). Preparation of polypyromellitimide.
52. No. 3,037,966 (1962). Preparation of polyimides with aliphatic groups in the chains.
53. No. 3,075,784 (1963). Preparation of conducting polyimides by treatment of polyamido-acids with organic acid salts.
54. No. 3,073,785 (1963). Preparation of current-conducting polyimides by chelation of polyamido-acids.
55. No. 3,168,417 (1965). Improvement of properties of Teflon-insulated electric wires by coating with polyimide.
56. No. 3,179,614 (1965). Preparation of polyamido-acids and polyamido-acid-based compositions.
57. No. 3,179,630 (1965). Preparation of polyamido-acid and polyimide films and fibers.
58. No. 3,179,631 (1965). Preparation of polyimide press powders.
59. No. 3,179,632 (1965). Preparation of polyimides by treating polyamido-acids with aromatic monocarboxylic acid anhydrides.
60. No. 3,179,634 (1965). Preparation of aromatic polyimides.

61. No. 3,179,635 (1965). Preparation of linear polyimides by modification of
     polyamides,
62. No. 3,207,728 (1965). Preparation of polyimides from thiophenetetracarboxylic
     acid and diamines.
63. No. 3,234,181 (1966). Preparation of aromatic polyimides.
64. No. 3,242,128 (1966). Preparation of coatings from compositions which contain
     a polyamido-acid.
65. No. 3,242,136 (1966). Preparation of ammonium salts of aromatic polyamido-
     acids and the corresponding polyimides.
66. No. 3,249,561 (1966). Preparation of polyimide plastic foams.
67. No. 3,249,588 (1966). Preparation of polyimide powder.

FRENCH PATENTS

68. No. 1,356,613 (1964). Preparation of solid resin powder from a polyamido-acid.
69. No. 1,360,488 (1964). Preparation of polyimide varnishes.
70. No. 1,361,712 (1964). Preparation of solid jet fuel from various resins, including
     polyimide resins.
71. No. 1,365,545 (1964). Preparation of polyimide press powders.
72. No. 1,368,741 (1964). Preparation of polyester-imide resin.
73. No. 1,373,383 (1964). Preparation of aromatic polyimides from dianhydrides
     and diesters of tetracarboxylic acids for use as electroinsulating coatings.
74. No. 1,379,219 (1964). Preparation of polyimide-based abrasive materials.
75. No. 1,386,258 (1965). Preparation of two-layered (polyimide and polyester)
     enamel insulation on electric wire coils.
76. No. 1,386,617 (1965). Improvement of properties of polyimides prepared from
     trimellitic anhydride.
77. No. 1,391,834 (1965). Preparation of polyester-polyimide composition for
     insulating electric wires.
78. No. 1,393,312 (1965). Preparation of polyimide with dispersed polytetrafluoro-
     ethylene particles.
79. No. 1,399,077 (1965). Preparation of aromatic polyimides by thermal treatment
     of polyamido-esters.
80. No. 1,399,078 (1965). Preparation of aromatic polyimides with a fluorinated
     or chlorinated diamine component.
81. No. 1,410,492 (1965). Preparation of polyimides from dianhydrides of bicyclic
     tetracarboxylic acids and primary aliphatic or aromatic diamines.
82. No. 1,414,526 (1965). Preparation of coatings from linear aromatic polyimides.
83. No. 1,421,681 (1965). Preparation of amido-imide polymers.
84. No. 1,422,458 (1965). Preparation of starting materials for the synthesis of
     polyimides.
85. No. 1,422,925 (1966). Preparation of polyester-imides.
86. No. 1,424,046 (1966). Preparation of a polyimide powder.
87. No. 1,427,087 (1966). Preparation of new dicarboxylic acids in the pyromelliti-
     mide series and their polycondensation with diamines and diols.

88. No.1,427,126 (1966). Preparation of polyimide copolymers and their utilization.
89. No.1,429,425 (1966). Preparation of polyamido-amines and polyimides.
90. No.1,429,946 (1966). Modification of polyimides by the introduction of silicon compounds.
91. No.1,430,968 (1966). Preparation of polyester-imides.
92. No.1,432,038 (1966). Preparation of fiber from polypyromellitimide.
93. No.1,436,595 (1966). One-stage preparation of polyimide objects.
94. No.1,441,930 (1966). Preparation of polyimides soluble in organic solvents.
95. No.1,446,178 (1966). Preparation of pressed objects from polyimides.
96. No.1,449,695 (1966). Preparation of new polyimide-based plastic materials

JAPANESE PATENTS

97. No.3432 (1965). Preparation of polyimide plastic foams.
98. No.3433 (1965). Preparation of polyimide plastic foams.
99. No.10,715 (1965). Improvement of formability of polyimide composition by the addition of halogenated hydrocarbon resin.

SOVIET REGISTRATION OF INVENTIONS

100. No.170,677 (1965). Preparation of polyimide by γ-irradiation of N-carbamyl-maleic acid imide.
101. No.171,552 (1965). Preparation of polyimides from 4,4'-diaminodiphenyl phthalide.
102. No.173,930 (1965). Increasing the molecular weight of polypyromellitamido acids.
103. No.173,931 (1965). Preparation of polyimides from butane-tetracarboxylic acid dianhydride.
104. No.180,703 (1966). Manufacture of varnish-film condensers with a polyimide dielectric.
105. No.188,005 (1966). Preparation of polyimides from bis-(4-aminophenoxy)-benzene.

Appendix II

REAGENTS EMPLOYED IN POLYIMIDE SYNTHESIS

1. DIAMINES

Aliphatic diamines: hexamethylene diamine; heptamethylene diamine; octamethylene diamine; nonamethylene diamine; decamethylene diamine; 2,11-diaminododecane; di-(p-aminocyclohexyl)-methane; 3-methylheptamethylene diamine; 4,4'-dimethylheptamethylene diamine; 1,2-bis-(3-aminopropoxy)-ethane; 2,2'-dimethylpropylene diamine; 2,5-dimethylheptamethylene diamine; 3-methyl-octamethylene diamine; 5-methylnonamethylene diamine; 1,10-diamino-1,10-dimethyldecane; 1,12-diaminooctadecane; $H_2N-(CH_2)_3-O-(CH_2)_2-O-(CH_2)_2-NH_2$; $H_2N-(CH_2)_3-S-(CH_2)_3-NH_2$; $H_2N-(CH_2)_3-N(CH_3)-(CH_2)_3-NH_2$.

Aromatic diamines: m- and p-phenynelene diamines; m- and p-xylylene diamines; benzidine; 3,3'-dimethoxybenzidine; dianisidine; 4,4'-diaminodiphenyl ether; 4,4'-diaminodiphenylpropane; 4,4'-diaminodiphenylmethane; bis-(4-amino-phenyl)-N-methylamine; 2,4-bis-(β-amino-tertbutyl)toluene; bis-(1-methyl-5-aminophenyl)-benzene; bis-(p-β-methyl-δ-aminophenyl)-benzene; p-bis-(2-methyl-4-aminoamyl)-benzene; p-bis-(1,1'-dimethyl-5-aminoamyl)-benzene; 4,4'-diaminodiphenyl sulfide; 4,4'-diaminodiphenyl sulfone; 3,3'-diaminodiphenyl sulfone; 1,5-diaminonaphthalene; 1,4-bis-(p-aminophenoxy)-benzene; 1,3-bis-(p-aminophenoxy)-benzene; diamines of the formula $H_2N-\langle\overline{}\rangle-X-\langle\overline{}\rangle-NH_2$.

where

$$X = -\overset{\overset{\textstyle R}{|}}{\underset{\underset{\textstyle R}{|}}{Si}}-; \quad -O-\overset{\overset{\textstyle R}{|}}{\underset{\underset{\textstyle R}{|}}{Si}}-O-; \quad -\overset{\overset{\textstyle R}{|}}{\underset{\underset{\textstyle O}{\parallel}}{P}}-; \quad -O-\overset{\overset{\textstyle R}{|}}{\underset{\underset{\textstyle O}{\parallel}}{P}}-O-.$$

2. DIANHYDRIDES

Dianhydrides of the following acids are employed.

Aliphatic acids: ethylenetetracarboxylic; 1,2,3,4-butane-tetracarboxylic; cyclopentane-1,2,3,4-tetracarboxylic.

Aromatic acids: pyromellitic acid and the following other tetracarboxylic acids: bis-(2,3-dicarboxyphenyl)methane; bis-(3,4-dicarboxyphenyl)methane; 1,1-bis-(2,3-dicarboxyphenyl)-ethane; 1,1-bis-(3,4-dicarboxyphenyl)ethane; 2,2-bis-(3,4-dicarboxyphenyl)propane; 2,2-bis-(2,3-dicarboxyphenyl)propane; bis-(3,4-dicarboxyphenyl) oxide; bis-(3,4-dicarboxyphenyl) sulfone; 3,3',4,4'-, 2,3,2', 3'- and 2,3,3',4'-benzophenonetetracarboxylic; 2,3,2',3'- and 3,3'4,4'-diphenyl-tetracarboxylic; 1,2,4,5-, 2,3,6,7-, 1,2,5,6- and 1,4,5,8-naphthalenetetra-carboxylic; 1,4,5,8-decahydronaphthalenetetracarboxylic; 2,6-dichloronaphthalene-1,4,5,8-tetracarboxylic; 2,7-dichloronaphthalene-1,4,5,8-tetracarboxylic; 2,3,6,7-tetrachloronaphthalene and 1,4,5,8-naphthalenetetracarboxylic; perylene-3,4,9,10-tetracarboxylic; phenanthrene-1,8,9,10-tetracarboxylic.

Heterocyclic tetracarboxylic acids: 2,3,4,5-thiophene-tetracarboxylic; 2,3,4,5-pyrrolidinetetracarboxylic; 2,3,5,6-pyridinetetracarboxylic.

3. SOLVENTS

N,N-dimethylacetamide; tetramethyl sulfone; N,N-dimethyl-formamide; dimethyl tetramethylene sulfone; tetramethylurea; N,N-diethylformamide; dimethyl sulfoxide; N,N-dimethoxyacetamide; N-methyl-2-pyrrolidone; N-methyl-caprolactam; pyridine; formamide; dimethyl sulfone; N-methylformamide; hexamethylphosphoramide; N-acetyl-2-pyrrolidone.

4. ANHYDRIDES

Chemical imidization is carried out with the aid of monocarboxylic acid anhydrides: acetic, propionic, butyric, valeric, benzoic, m-and p-ethylbenzoic, p-propylbenzoic, p-isopropyl-benzoic, o-, m- and p-nitrobenzoic, o-, m- and p-halobenzoic, α-naphthoic, β-naphthoic.

INFRARED ABSORPTION SPECTRA OF A NUMBER OF POLYIMIDES*

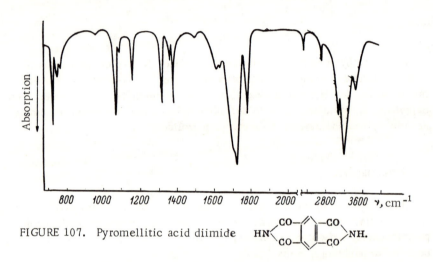

FIGURE 107. Pyromellitic acid diimide

*   The spectra represented in this Appendix were taken by E. F. Fedorova and
     E. I. Pokrovskii, who used a Japanese DS-301 spectrophotometer, and by the authors
     of this book, who used an IKS-14 spectrophotometer of Soviet manufacture. Some
     of the spectra given in the Appendix are based on literature data.

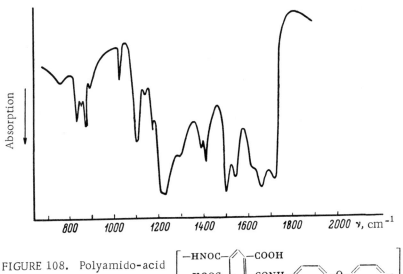

FIGURE 108. Polyamido-acid $\left[\begin{array}{c} -HNOC-\bigcirc-COOH \\ HOOC-\bigcirc-CONH-\bigcirc-O-\bigcirc- \end{array}\right]_n$

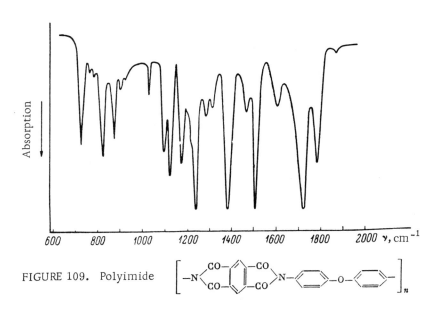

FIGURE 109. Polyimide $\left[-N\begin{array}{c} CO-\bigcirc-CO \\ CO-\bigcirc-CO \end{array}N-\bigcirc-O-\bigcirc-\right]_n$

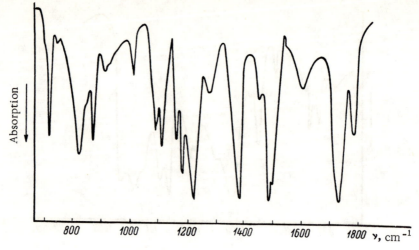

FIGURE 110. Polyimide $\left[-N\begin{matrix} CO \\ CO \end{matrix} \right.$ ... $\left. \right]_n$.

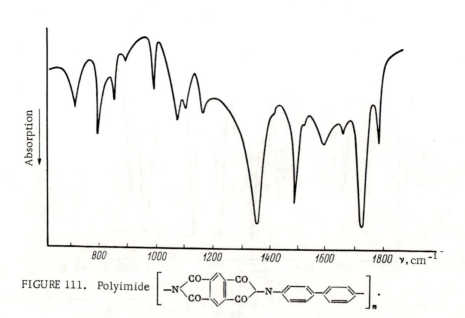

FIGURE 111. Polyimide $\left[-N\begin{matrix} CO \\ CO \end{matrix} \right.$ ... $\left. \right]_n$.

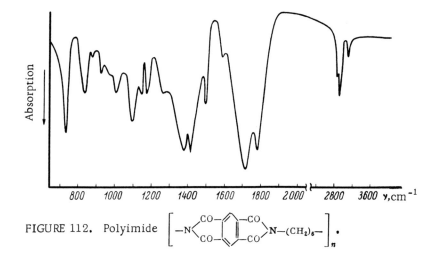

FIGURE 112. Polyimide $\left[-N \begin{smallmatrix} CO \\ CO \end{smallmatrix} \bigcirc \begin{smallmatrix} CO \\ CO \end{smallmatrix} N-(CH_2)_6- \right]_n$.

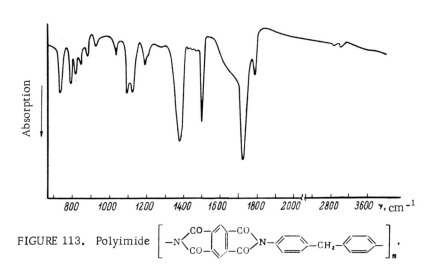

FIGURE 113. Polyimide $\left[-N \begin{smallmatrix} CO \\ CO \end{smallmatrix} \bigcirc \begin{smallmatrix} CO \\ CO \end{smallmatrix} N-\bigcirc-CH_2-\bigcirc- \right]_n$.

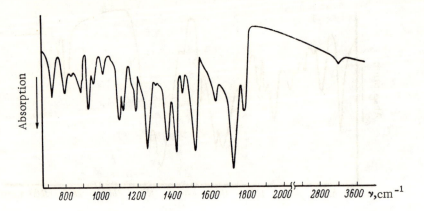

FIGURE 114. Polyimide

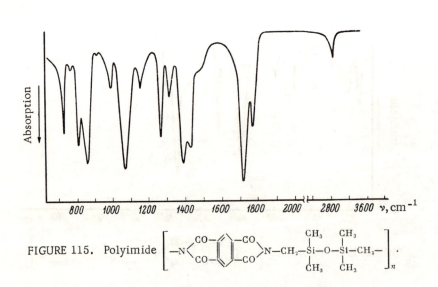

FIGURE 115. Polyimide

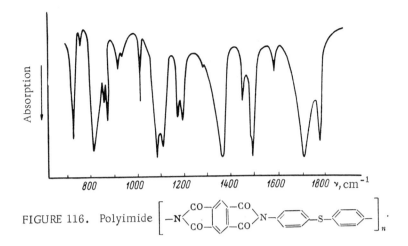

FIGURE 116. Polyimide $\left[-N\begin{smallmatrix} CO \\ CO \end{smallmatrix} \right\rangle N - \text{—} S \text{—} \right]_n$.

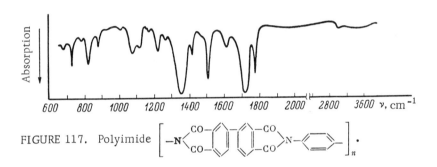

FIGURE 117. Polyimide $\left[-N\begin{smallmatrix} CO \\ CO \end{smallmatrix} \right\rangle N - \text{—} \right]_n$.

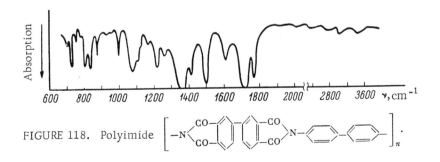

FIGURE 118. Polyimide $\left[-N\begin{smallmatrix} CO \\ CO \end{smallmatrix} \right\rangle N - \text{—} \right]_n$.

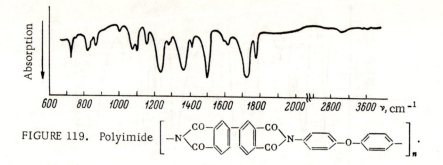

FIGURE 119. Polyimide $\left[-N\begin{smallmatrix}CO\\CO\end{smallmatrix} \right]_n$.

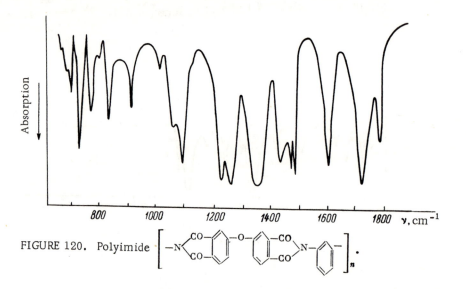

FIGURE 120. Polyimide $\left[-N\begin{smallmatrix}CO\\CO\end{smallmatrix} \right]_n$.

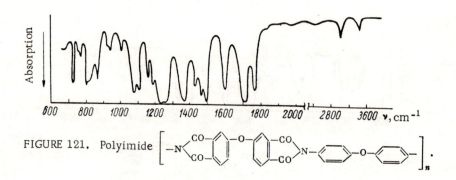

FIGURE 121. Polyimide $\left[-N\begin{smallmatrix}CO\\CO\end{smallmatrix} \right]_n$.

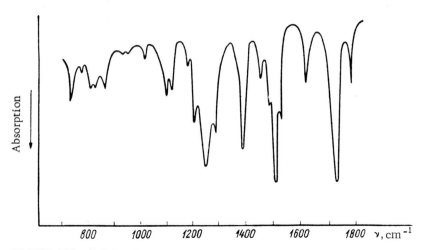

FIGURE 122. Polyimide

$$\left[-N\begin{matrix}CO\\CO\end{matrix}\bigcirc-O-\bigcirc\begin{matrix}CO\\CO\end{matrix}N-\bigcirc-O-\bigcirc-O-\bigcirc-\right]_n.$$

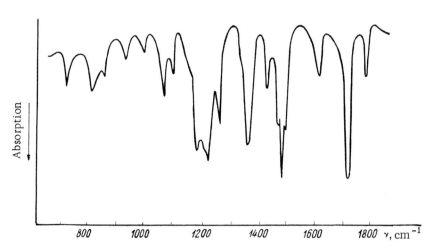

FIGURE 123. Polyimide

$$\left[-N\begin{matrix}CO\\CO\end{matrix}\bigcirc-O-\bigcirc-O-\bigcirc\begin{matrix}CO\\CO\end{matrix}N-\bigcirc-O-\bigcirc-O-\bigcirc-\right]_n.$$

161

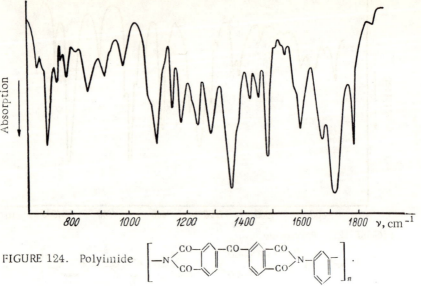

FIGURE 124. Polyimide $\left[-N \begin{array}{c} CO \\ CO \end{array} \bigcirc -CO- \bigcirc \begin{array}{c} CO \\ CO \end{array} N- \bigcirc - \right]_n$.

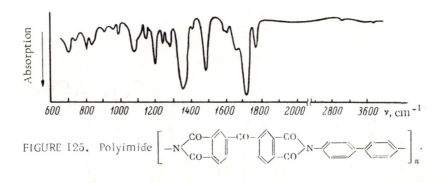

FIGURE 125. Polyimide $\left[-N \begin{array}{c} CO \\ CO \end{array} \bigcirc -CO- \bigcirc \begin{array}{c} CO \\ CO \end{array} N- \bigcirc - \bigcirc \right]_n$.

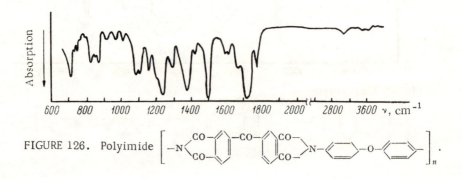

FIGURE 126. Polyimide $\left[-N \begin{array}{c} CO \\ CO \end{array} \bigcirc -CO- \bigcirc \begin{array}{c} CO \\ CO \end{array} N- \bigcirc -O- \bigcirc \right]_n$.